Springer Desktop Editions in Chemistry

T0223330

L. Brandsma, S.F. Vasilevsky,
H.D. Verkruijsse
Application of Transition Metal Catalysts
in Organic Synthesis
ISBN 3-540-65550-6

H. Driguez, J. Thiem (Eds.)
Glycoscience, Synthesis of Oligosaccharides
and Glycoconjugates
ISBN 3-540-65557-3

H. Driguez, J. Thiem (Eds.)
Glycoscience, Synthesis of Substrate Analogs
and Mimetics
ISBN 3-540-65546-8

K. Faber (Ed.)
Biotransformations
ISBN 3-540-66949-3

W.-D. Fessner (Ed.)
Biocatalysis, From Discovery to Application
ISBN 3-540-66970-1

S. Grabley, R. Thiericke (Eds.)
Drug Discovery from Nature
ISBN 3-540-66947-7

H.A.O. Hill, P.J. Sadler, A.J. Thomson (Eds.)
Metal Sites in Proteins and Models,
Iron Centres
ISBN 3-540- 65552-2

H.A.O. Hill, P.J. Sadler, A.J. Thomson (Eds.)
Metal Sites in Proteins and Models,
Phosphatases, Lewis Acids and Vanadium
ISBN 3-540-65553-0

H.A.O. Hill, P.J. Sadler, A.J. Thomson (Eds.)
Metal Sites in Proteins and Models,
Redox Centres
ISBN 3-540-65556-5

F.J. Leeper, J.C. Vederas (Eds.)
Biosynthesis, Polyketides and Vitamins
ISBN 3-540-66969-8

A. Manz, H. Becker (Eds.)
Microsystem Technology
in Chemistry and Life Sciences
ISBN 3-540-65555-7

P. Metz (Ed.)
Stereoselective Heterocyclic Synthesis
ISBN 3-540-65554-9

H. Pasch, B. Trathnigg
HPLC of Polymers
ISBN 3-540-65551-4

J. Rohr (Ed.)
Bioorganic Chemistry, Deoxysugars,
Polyketides and Related Classes: Synthesis,
Biosynthesis, Enzymes
ISBN 3-540-66971-X

T. Scheper (Ed.)
New Enzymes for Organic Synthesis,
Screening, Supply and Engineering
ISBN 3-540-65549-2

F.P. Schmidtchen (Ed.)
Bioorganic Chemistry,
Models and Applications
ISBN 3-540-66978-7

Springer

Berlin
Heidelberg
New York
Barcelona
Hong Kong
London
Milan
Paris
Singapore
Tokyo

J. Rohr (Ed.)

Bioorganic Chemistry
Deoxysugars, Polyketides and Related Classes: Synthesis, Biosynthesis, Enzymes

 Springer

Dr. Jürgen Rohr
Institut für Organische Chemie
der Georg-August-Universität
Tammannstr. 2
37077 Göttingen, Germany
E-mail: jrohr@gwdg.de

Description of the Series

The Springer Desktop Editions in Chemistry is a paberback series that offers selected thematic volumes from Springer chemistry series to graduate students and individual scientists in industry and academia at very affordable prices. Each volume presents an area of high current interest to a broad non-specialist audience, starting at the graduate student level.

Formerly published as hardcover edition in the review series
Topics in Current Chemistry (Vol. 188) ISBN 3-540-62488-0

Cataloging-in-Publication Data applied for

ISBN 3-540-66971-X
Springer-Verlag Berlin Heidelberg New York

Die Deutsche Bibliothek - CIP-Einheitsaufnahme
Bioorganic chemistry: deoxysugars, polyketides and related classes: synthesis, biosynthesis, enzymes / J.Rohr (ed.) - Berlin; Heidelberg; New York; Barcelona; Hong Kong; London; Milan; Paris; Singapore; Tokyo: Springer, 2000
(Springer desktop editions in chemistry)
ISBN 3-540-66971-X

Springer-Verlag is a company in the specialist publishing group BertelsmannSpringer
© Springer-Verlag Berlin Heidelberg 2000
Printed in Germany

Cover: design & production, Heidelberg
Typesetting: Fotosatz-Service Köhler OHG, Würzburg
Printed on acid-free paper   SPIN: 10720822   02/3020 hu - 5 4 3 2 1 0

Preface

"Bioorganic Chemistry" has developed into a major area of scientific investigation located in the force field at the interface of Organic Chemistry, Biochemistry, Medicinal Chemistry and Biology. This is a logical consequence of the general tendency of increasing overlap in all of the sciences, such that researchers can no longer isolate themselves in a limited environment, in which they, for example, continue to develop new methods. The question how to create an alternative or even better method has always to be followed by the question "why?", and it is certainly not good enough anymore to address this by a superficial one-sentence introduction. This naturally leads to interfaculty collaborations which expand the horizon of all those involved.

Novel scientific contributions in the broad field of Bioorganic Chemistry can be found not only in typical broad-based scientific and chemical or organic chemical journals, but also in numerous specialized journals dealing with synthetic organic chemistry, natural product chemistry, phytochemistry, biochemistry, bioconjugate chemistry, medicinal chemistry, microbiology, molecular biology, etc. Thus it is harder for anybody working or interested in this field to maintain an overview, and to "see" novel and promising tendencies. Advanced textbooks and monographs are useful introductions, but often cover only one aspect of such a broad field (e.g., biomimetics or biosynthesis or natural products) and cannot be expected to present the forefront of any given research area. Comprehensive review articles, on the other hand, can fill the gap and give the interested reader a comparably short and easy introduction or a progress report.

This book contains three comprehensive review articles about special, current topics in the broad field of Bioorganic Chemistry. However, they are interrelated and complement one another nicely. In all three contributions, the general theme "biosynthesis" plays a prominent or even dominating role. Furthermore, synthetic methods and strategies for important classes of natural products are highlighted in general, with special emphasis on one of the most varied (with regard to the structural diversity and biological activity) and most abundant groups of natural products, the polyketides. The review articles on deoxysugars and on non-template based multienzyme systems are the most topical and comprehensive ones written to date; the review about angucyclines, the largest subgroup of aromatic polyketides, contains the most comprehensive presentation of syntheses in this field and indicates the direction of current developments.

Although this book is the second volume under the heading "Bioorganic Chemistry", it is much more than a sequel to the previous volume; it contains other important facets of this broad field of interdisciplinary science. The editor whishes to express his thanks to all contributors and critical reviewers of the articles who have made this book possible.

Göttingen, January 1997   Jürgen Rohr

Table of Contents

Topics in Current Chemistry
Now Also Available Electronically

For all customers with a standing order for **Topics in Current Chemistry** we offer the electronic form via LINK **free of charge**. You will receive a password for free access to the full articles.
Please register at: **http://link.springer.de/series/tcc/reg_form.htm**

If you do not have a standing order you can nevertheless browse through the table of contents of the volumes and the abstracts of each article at:
http://link.springer.de/series/tcc

There you will also find information about the
- Editorial Board
- Aims and Scope
- Instructions for Authors

Chemical and Biochemical Aspects of Deoxysugars and Deoxysugar Oligosaccharides*

Andreas Kirschning[1] · Andreas F.-W. Bechthold[2] · Jürgen Rohr[3]

[1] Institut für Organische Chemie, Technische Universität Clausthal, Leibnizstraße 6, 38678 Clausthal-Zellerfeld, Germany. *E-mail: andreas.kirschning@tu-clausthal.de*
[2] Institut für Pharmazeutische Biologie, Universität Tübingen, Auf der Morgenstelle 8, 72076 Tübingen, Germany. *E-mail: bechthold@uni-tuebingen.de*
[3] Institut für Organische Chemie, Universität Göttingen, Tammannstraße 2, 37077 Göttingen, Germany. *E-mail: jrohr@gwdg.de*

Deoxysugars and the oligosaccharides derived from them belong to one of the most important, but often most neglected, group of biological compounds. As a result of important recent progress in their chemical synthesis as well as insights into their various biochemical aspects, one can expect a major boom in the amount of research devoted to this promising field. In this chapter, the most significant aspects of the biochemistry and chemistry of deoxysugars and deoxysugar-containing oligosaccharides are discussed: (1) biosynthetic, biochemical, and molecular biological studies, including deoxygenation mechanisms, glycosyl transfer, and genes involved in biosynthesis; (2) recent experiments aimed at determining how mono- or oligosaccharide moieties of bioactive natural products contribute to biological activity, i.e., the mode of action of these compounds; and (3) synthetic aspects such as chemical glycosidation and enzymatic methods for the construction of deoxysugar oligosaccharides. The information presented here is based on literature published until mid-1996.

Keywords: Deoxysugars, deoxysugar oligosaccharides, biosynthesis, deoxygenation mechanisms, glycosyl transfer, combinatorial biosynthesis, genetics, enzymes, antibiotics, antitumor compounds, bioactive natural products, hybrid natural products, DNA interactions, glycosidation methods, oligosaccharide syntheses, enzymatic syntheses, enzymatic glycosidations.

Table of Contents

---

* Dedicated to Heinz G. Floss.

Topics in Current Chemistry, Vol. 188
© Springer-Verlag Berlin Heidelberg 1997

1
Introduction

Carbohydrates are the most abundant group of natural products and the role of sugars and deoxysugars as energy and biosynthetic resources (glycolysis, pentose phosphate cycle, shikimate pathway, etc.), "energy storage devices" (photosynthesis) and key structural elements in the formation of biological backbones (2-deoxyribose for DNA or N-acetylglucosamine for murein), is general knowledge [1–6]. Carbohydrates and carbohydrate-containing structural moieties are also involved in more active biochemical and bioorganic processes. They are important elements of recognition and specificity in cell-cell interactions [7, 8], for example, as cell surface oligosaccharides [9, 10], e.g., tumor associated antigens [11, 12], lectins [13, 14], glycoproteins, glycolipids, and immunodeterminants. They also play a part in the mode of action of many drugs as they contribute to a variety of processes, including active trans-membrane transport [15, 16], stabilization of protein folding [17], and enzyme inhibition [18, 19].

Deoxysugars are frequently found, either as single structural elements or, more frequently, as components of oligosaccharides, in antibiotics and anti-cancer agents such as anthracyclines, angucyclines, aureolic acid antibiotics, avermectins, cardiac glycosides, enediynes, macrolides, pluramycins and others (for some examples, see Sects. 2, 4). A large number of the so-called bioactive carbohydrates in lipopolysaccharides as well as several antibiotics are deoxy-sugars [1], especially if one also takes amino sugars, i.e., those carbohydrates in which an oxygen has been replaced by a nitrogen, into consideration. In this article deoxysugars are defined as all carbohydrates in which one or more of the normally (e.g., in glucose) occurring oxygen atoms are deleted (i.e., replaced by hydrogen) or replaced by any other heteroatom or heteroatomic group, such as sulfur (thiosugars), halogen, nitrogen (aminosugars) or NO_x (nitro- and nitro-sosugars) (Fig. 1a).

More than 13 years ago, Kennedy and White, in their excellent monography on bioactive carbohydrates [1], listed over 100 different naturally occurring deoxymonosaccharides or monosaccharide derivatives present, e.g., in lipo-polysaccharides or antibiotics. Several more have been discovered since then, for

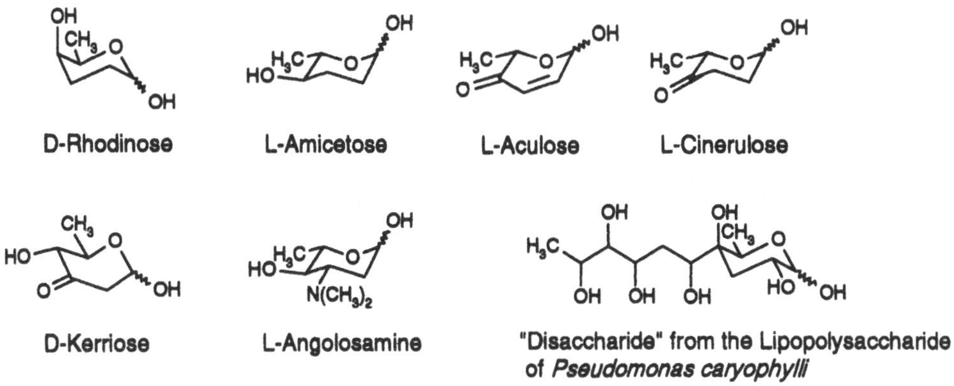

D-Rhodinose   L-Amicetose   L-Aculose   L-Cinerulose

D-Kerriose   L-Angolosamine   "Disaccharide" from the Lipopolysaccharide of *Pseudomonas caryophylli*

Fig. 1a

example, D-rhodinose, L-amicetose, L-aculose, L-cinerulose, D-kerriose, and L-angolosamine (for additional examples see [19, 20] and the chapter on the angucycline group of antibiotics). Nevertheless, little is known about the bioactivity of the deoxysugars themselves or that of deoxysugar-containing oligosaccharides. As components of bioactive natural products, their contribution to the mechanism of action is also poorly understood [21]. Furthermore, with the exception of deoxyribose [22] and the 3,6-dideoxysugar ascarylose (see below), there is little information available regarding the biosynthesis of single deoxysugars or oligosaccharides. Finally, it should also be noted that deoxysugar moieties pose unique difficulties in chemical syntheses, in particular in glycosidation reactions of 2-deoxysugars, in which a control element in the neighboring 2-position is missing (see Sect. 5).

The goal of this chapter is to provide a brief insight into the chemistry and biochemistry of deoxysugars, starting with an overview of the biosynthetic steps (Sect. 2) leading to formation of single bioactive deoxysugars and oligosaccharides and a discussion of the genes that code for the enzymes controlling these biosynthetic steps (Sect. 3). This is followed by a discussion of several important deoxysugars and oligosaccharides and their function in some of the bioactive drugs containing these elements. (Sect. 4). Finally, recent developments in oligosaccharide synthesis, including enzymatic methods, with particular emphasis on 2-deoxysugars, will be presented, further exemplified by selected highlights in the preparation of complex deoxygenated oligosaccharides (Sect. 5).

2
Biosyntheses of Deoxysugars and Deoxysugar-Containing Oligosaccharide Moieties of Natural Products

This section reviews some of the biosynthetic studies carried out on biologically active natural products possessing a deoxysugar-containing carbohydrate moiety in which biosynthesis of the latter was examined, at least to some extent. Although the contribution of deoxysugars to biological activity is well recognized, our knowledge concerning the biosynthesis of deoxysugars and their subsequent assembly into oligosaccharide moieties is still very limited [21, 23 – 25]. While many schemes have become generally accepted, they nonetheless remain highly speculative. However, important contributions have come from biochemical work (see below), and hopefully, there can be more expected from the analysis of genes that code for enzymes involved in deoxysugar biosynthesis (see also Sect. 3); several details have already been discussed in the excellent review of Liu and Thorson [21].

2.1
Deoxygenation Mechanisms

Nearly everything organic in nature starts with glucose, since the photosynthetic system of plants yields this sugar as a result of conversion of solar light energy into chemical "storage" energy. Although the deoxygenation of glucose into deoxysugars may not have been an attractive field of research, the studies nevertheless yielded insight into a variety of biosynthetic mechanisms as well as into the molecular machinery of numerous organisms. However, the broader question, how D-glucose loses oxygen in order to be transformed into a highly deoxygenated sugar (Scheme 1), e.g., D-amicetose (1, 2,3,6-trideoxy-D-*erythro*-hexopyranose), L-rhodinose (2, 2,3,6-trideoxy-L-*threo*-hexopyranose), L-daunosamine (3, 3-amino-2,3,6-trideoxy- L-*lyxo*-hexopyranose) or L-tolyposamine (4, 4-amino-2,3,4,6-tetradeoxy-L-*erythro*-hexopyranose) [1, 26, 27], is still largely unanswered. One major obstacle was the long-held dogma that incorporation experiments with putative intermediates were not possible (but see Sect. 2.3.3), since syntheses of the activated nucleoside diphosphate deoxysugars (NDP-deoxysugars), postulated as intermediates, are complicated. This is especially

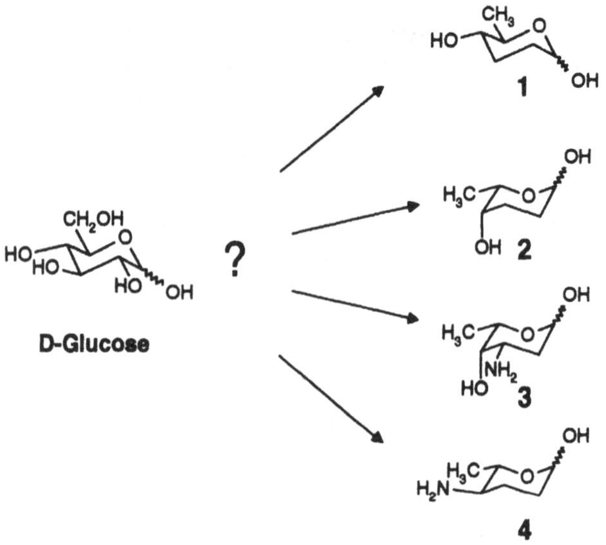

Scheme 1. Deoxygenation mechanisms of highly deoxygenated sugars are still widely unknown

true for those intermediates occurring further down in the deoxygenation cascade, for instance, when the 2-position is already deoxygenated, as will be further discussed in Sect. 5 [23, 28–32].

2.1.1
6-Deoxygenation

The formation of NDP-6-deoxy-4-hexuloses (Scheme 2), e.g., NDP-6-deoxy-D-*xylo*-4-hexulose (5, NDP-4-keto-6-deoxy-D-glucose), has been the subject of intensive research for more than three decades. An NDP-4-keto-6-deoxy intermediate was first suggested by Kornfeld and Glaser, in the formation of L-rhamnose from D-glucose, and shortly afterwards by Ginsburg, for the formation of L-fucose from D-mannose [33–35]. The mechanism of NDP-4-keto-6-deoxyglucose generation (Scheme 2) was studied using model systems and the isolated deoxythymidinediphosphate-D-glucose (dTDP-D-glucose) oxidoreductase [36–38]. The stereochemical course of the intramolecular *H*-shift was determined mainly by Floss and coworkers [39] (see also furtherwork based on these fundamental studies [40–43]), who used dTDP-(6R)- and -(6S)-[4-^{2}H,

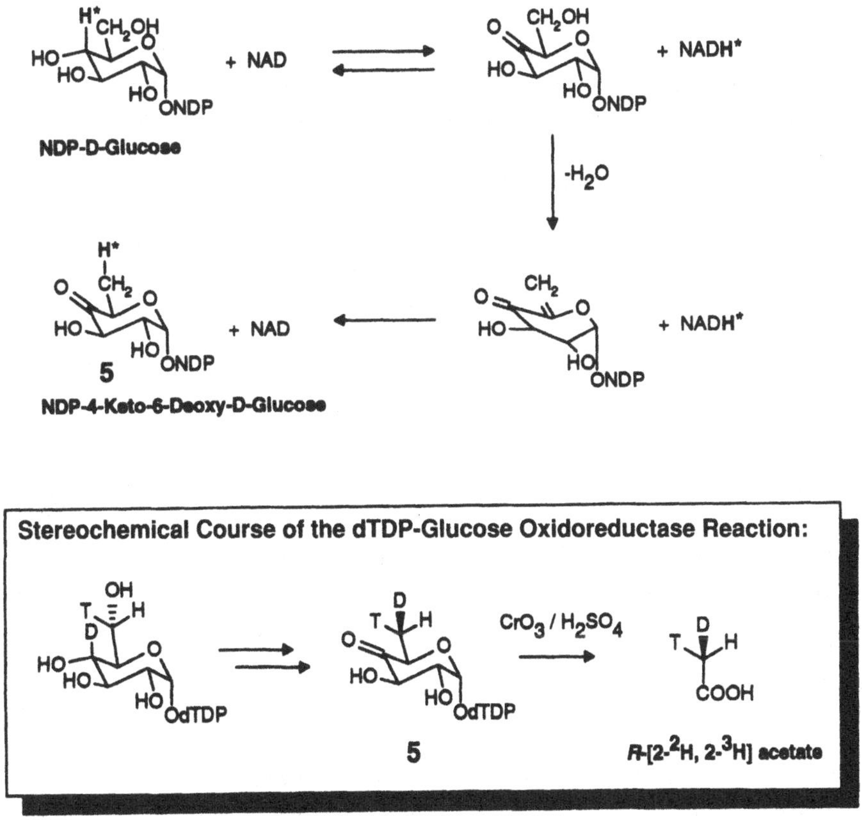

Scheme 2. Generation and stereochemical course of NDP-4-keto-6-deoxyglucose (5)

6-^{3}H]glucose as substrate [39] with the enzyme from E. coli. The configurational analysis (using the enzymatic method of Cornforth and Arigoni [44, 45]) of chiral acetic acid, obtained by Kuhn-Roth oxidation of the resulting 6-deoxy-sugar nucleotide, confirmed that the H-shift from C-4 to C-6 occurs intramolecularly and showed that the migrating hydrogen replaces the 6-OH group with inversion of configuration. Later, this was also shown for the GDP-D-mannose dehydratase [46 a] (the "mannose analog" of NDP-glucose oxidoreductase) from an unidentified soil bacterium as well as for the NDP-glucose oxidoreductase reactions in various *Streptomyces* [41–43] and in *Yersinia pseudotuberculosis* [40]. In contrast to other dehydratases, which possess a tightly bound NAD$^+$, the purified CDP-D-glucose-4,6-dehydratase from *Yersinia pseudotuberculosis* exhibits an absolute NAD$^+$ requirement for activity [40]. The substrate specificity of a dehydratase from *Salmonella typhimurium* LT2 expressed in *E. coli* has already been investigated (see Sect. 5.2.1.2, Scheme 48). These results suggest that members of this class of enzymes derive from a common ancestor whose mechanism and therefore stereochemical course has been preserved throughout the evolution of the diverse descendants [40, 46 b].

2.1.2
3-Deoxygenations

3,6-Dideoxyhexoses are frequently found in the lipopolysaccharide components of the cell walls of gram-negative colon bacteria [1]. These compounds have attracted attention due to their immunogenicity. Extensive biosynthetic investigations on L-ascarylose were carried out to establish the general pathway leading to this class of deoxysugars. A component of this pathway, 3-deoxygenation has, in turn, been well-studied mechanistically [40, 47–54]. Removal of the 3-oxygen with retention of configuration is catalyzed by two enzymes and begins with CDP-4-keto-6-deoxyglucose. First, a pyridoxaminephosphate (PMP)-linked dehydrase attacks the 4-keto group to yield the imine, followed by a 1,4-elimination of water. The resulting CDP-6-deoxy-$\Delta^{3,4}$-glucoseen intermediate is then reduced by an Fe-S cluster in the dehydrase (E1) and the oxidized Fe-S cluster in the dehydrase is then reduced by the [2Fe-2S]-containing flavoprotein (E3). This follows from the finding that E1 alone can catalyze the entire deoxygenation reaction in the presence of an alternative ē donor, e.g., diaphorase of the reductase component of methanemonooxygenase. Cleavage of the PMP-enzyme yields the 3,6-dideoxy-4-ketohexose **6** which finally is then epimerized and reduced to CDP-L-ascarylose **7** (Scheme 3).

Biosynthetic studies on the nucleoside antibiotic blasticidin S **8**, using [2, 3, 4, 6, 6-^{2}H$_5$]glucose, [3-^{2}H]-D-glucose, NADP^{2}H, generated in situ from [1,1-^{2}H$_2$]ethanol, and inhibitors of PMP-dependent transaminases revealed the deoxygenation sequence leading to the aminodeoxyhexuronic acid moiety **9** of **8** [55]. Similar to the biosynthetic sequence depicted in Scheme 3, an initial attack of the PMP cofactor at the postulated 4-ketoderivative induces elimination of first the 3- and then the 2-oxygen. The ^{2}H atoms of glucose remaining in **8** support the proposed reaction sequence (Scheme 4, see also the following section).

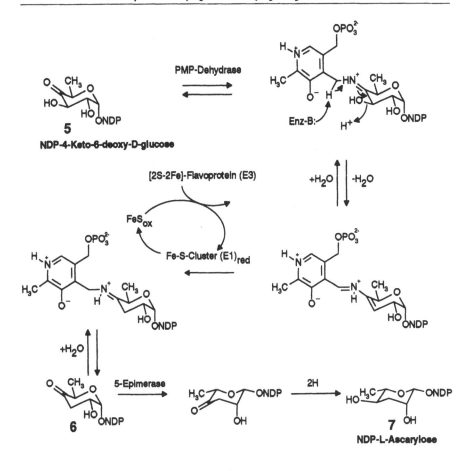

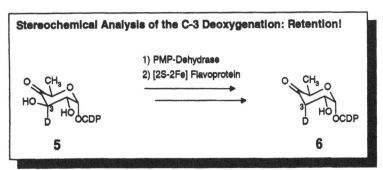

Scheme 3. 3-Deoxygenation of NDP-L-ascarylose (**7**) with a pyridoxamine phosphate (PMP)-dependent enzyme, and its stereochemical course

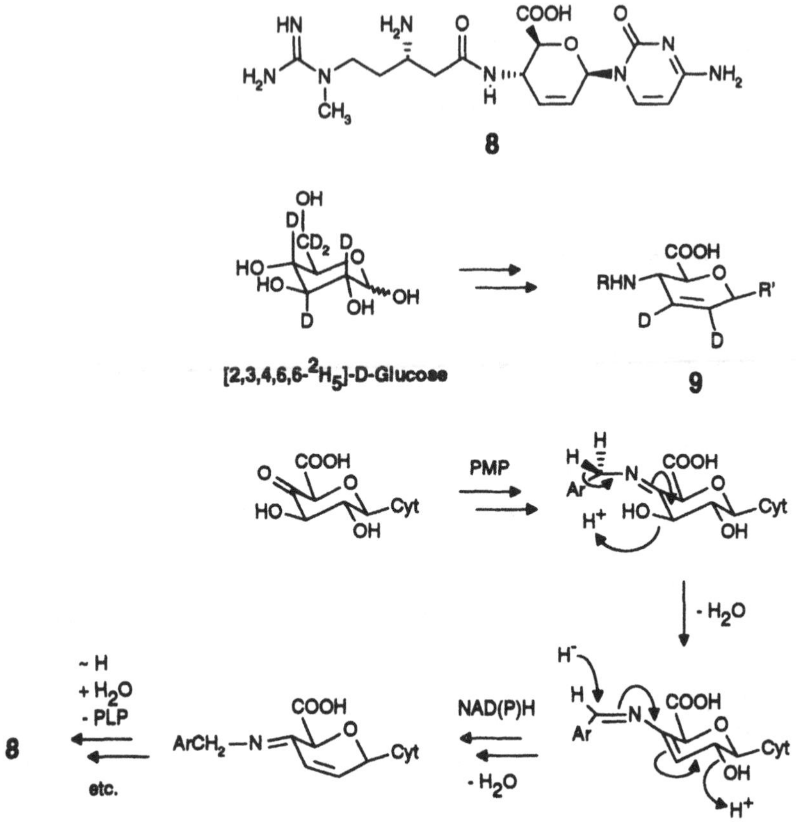

8

[2,3,4,6,6-^{2}H$_5$]-D-Glucose   **9**

Scheme 4. Deoxygenation of the nucleoside antibiotic blasticidin S (**8**). A PMP (pyridoxamine phosphate) may control both the 3- and the 2-deoxygenation (both under retention of protons) from the 4-keto group. PLP, pyridoxal phosphate

2.1.3
2-Deoxygenation

2,6-Dideoxysugars, sometimes along with 2,3,6-trideoxysugars, are frequently found in secondary metabolites of microorganisms, especially in those produced by the genus *Streptomyces*. The conclusion that the pathway leading to 2,6-dideoxysugars cannot be based on the 3,6-dideoxysugar biosynthetic pathway found in gram-negative colon bacteria (e.g., *Yersinia*, see Scheme 3) [53, 54] was based on two observations: (1) a comparison of the genes involved in ascarylose biosynthesis with those (see also Sect. 3) required for production of 2,6-dideoxy- and 4,6-dideoxysugar-containing antibiotics and (2) stereochemical arguments arising from incorporation experiments with stereospecifically isotope-labeled glucose derivatives. By extensive feeding experiments using ^{14}C-, ^{3}H- or ^{2}H-labeled glucose samples on the granaticin producer *Streptomyces violaceoruber* Tü22, biosynthesis of the 2,6-dideoxysugar in granaticin **10** was examined [42]. Again, the biosynthetic sequence is initiated by formation

of NDP-4-keto-6-deoxyhexose. To determine the stereochemical course of replacement of the 2-hydroxy group by hydrogen, two key experiments were
carried out. Feeding of $[3, 4-^{14}C_2, 2-^3H]$glucose showed that 2-H of glucose is
retained, and feeding of $[2-^2H]$glucose demonstrated that the 2-H of glucose
ends up in the pro-S 2'-position of granaticin (10), i.e., the hydroxy group at
C-2 in glucose is replaced by H with retention of configuration (Scheme 5a). Since feeding studies with $[3, 4-^{14}C_2, 3-^3H]$glucose revealed that the 3-H of glucose
is lost during the biosynthesis of granaticin, a mechanism similar to 3-deoxy-
genation [47, 48, 54] (see above) under control of a pyridoxamine cofactor-con-
taining enzyme was originally favored. This requires initial formation of a
3-keto intermediate, i.e., a shift from 4-keto-6-deoxy- to 3-keto-6-deoxyhexose.
By contrast, the 2-deoxygenations that occur during formation of the 2,6-
dideoxysugar (β-D-olivose) moieties of chlorothricin (11) by *Streptomyces antibioticus* Tü99 proceed with inversion of configuration at C-2 [43] (Scheme 5b)
when the 2-OH of glucose and dTDP-4-keto-6-deoxy-D-glucose is replaced by
hydrogen. Thus it is likely that more than one pathway is used to achieve the
2-deoxygenation and that, as mentioned above, a PMP-containing enzyme, as
found in the 3-deoxygenation, which has been proven to operate under retention

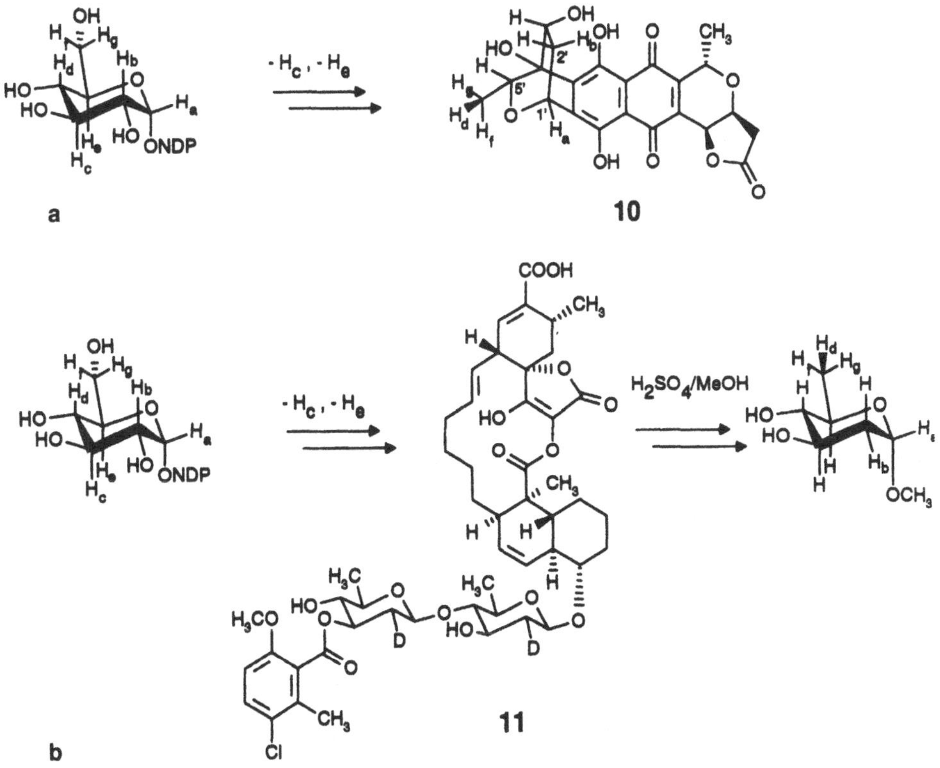

Scheme 5a, b. Deoxygenation steps in the biosynthesis of 2,6-dideoxysugars in granaticin (10)
and chlorothricin (11). Note that the 6-deoxygenation ocurs in both cases under inversion,
while the 2-deoxygenation occus under retention (in 10) or under inversion (in 11)

of configuration at the 3-position (see Scheme 3) may not always be involved in the 2-deoxygenation step. Again, the studies on the 2,3,4-trideoxyglucuronic acid moiety **9** in blasticidin S (**8**) [55] showed retention of ^{2}H in both the 2- and 3-positions, when [2, 3, 4, 6, 6-^{2}H$_5$]- and [3-^{2}H]-glucose were fed to *Streptomyces griseochromogenes*. This clearly excludes a 2- or 3-keto intermediate, as is considered necessary for a pyridoxamine cofactor-linked deoxygenase. Note that, in the case of the formation of **9**, the pyridoxamine cofactor is suggested to control both the 2- and the 3-deoxygenation from the 4-position (Scheme 4).

Feeding [1, 2, 3, 4, 5, 6, 6-^{2}H$_7$]glucose to the landomycin A (**12**) producer *Streptomyces cyanogenus* S-136 revealed that only the ^{2}H-atoms in the 1- and 6-positions are clearly retained in both the D-olivose and the L-rhodinose moieties [56]. This is indicated by ^{2}H-NMR analysis, performed on both landomycin A and the α-methyl olivoside isolated after methanolysis of **12**. Thus the ^{2}H atoms seem to be removed from the 2-, 3- and 4-positions of [U-^{2}H$_7$]glucose during biosyntheses of these 2,6-di- and 2,3,6-trideoxy-sugars (Scheme 6a). Feeding experiments using [1-^{13}C]glucose always yielded better incorporation into the L-rhodinose than into the D-olivose moieties of the hexadeoxysaccharide chain of landomycin A (**12**), regardless of the time period during which the feeding was performed [56, 57]. From this, it has to be concluded that the 2,6-dideoxysugar NDP-D-olivose **13** cannot be a biosynthetic precursor of the 2,3,6-trideoxysugar NDP-L-rhodinose **13b**, and the two pathways have to branch early, preferably as shown in Scheme 6b, or already at the NDP-4-keto-6-deoxy-D-glucose **5** step. The latter is also assumed to be the branch point in the biosynthesis of the two deoxysugars in erythromycin A (**14**), D-desosamine and L-mycarose (which will become L-cladinose through 3-O methylation at the very end of **14** biosynthesis, Scheme 7) [24, 53]. In sum-

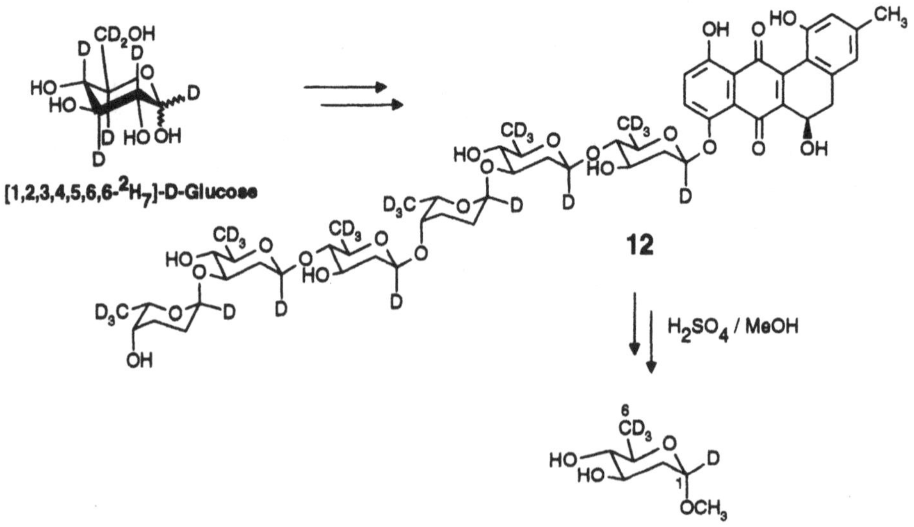

Scheme 6a. Feeding experiment with [U-^{2}H$_7$]glucose on landomycin A (**12**). Only the 1- and 6-positions show detectable amounts of ^{2}H label

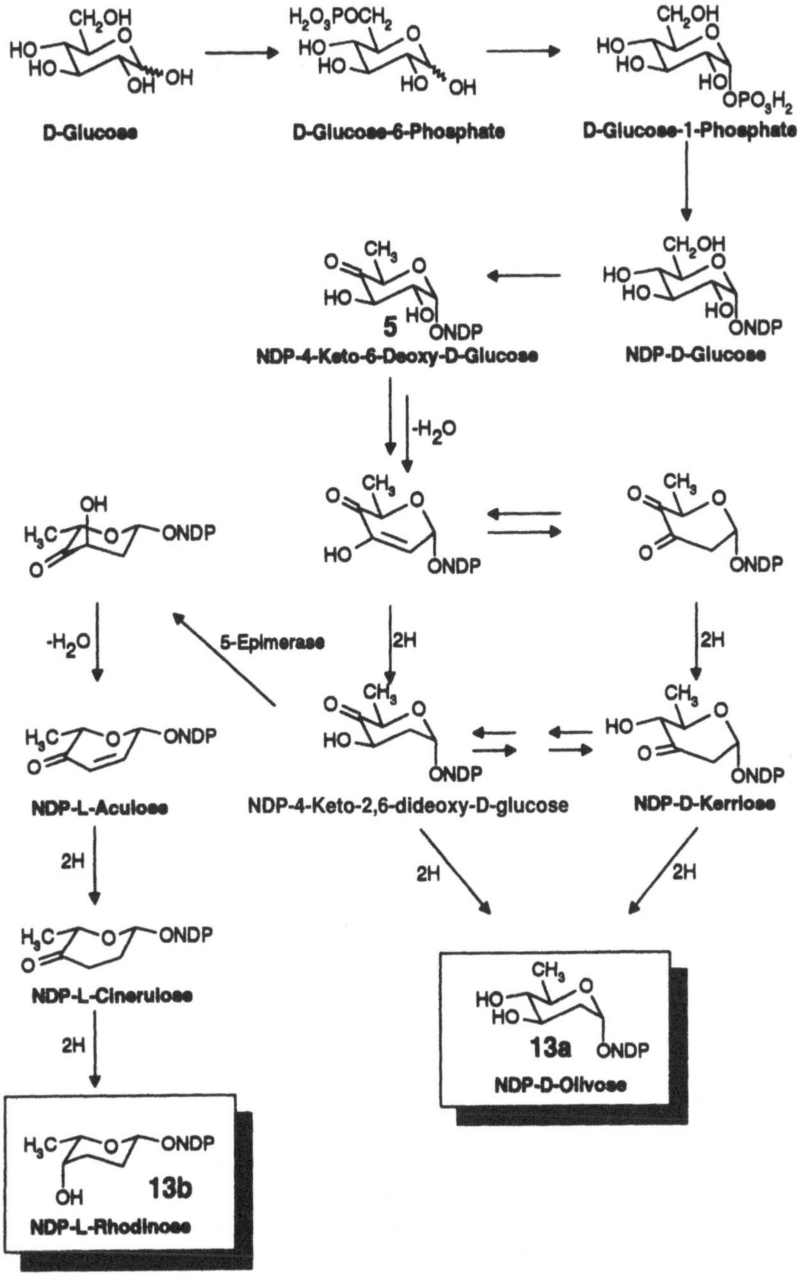

Scheme 6b. Favored hypothetical deoxygenation pathways to the deoxysugars of landomycin A, NDP-D-olivose (**13a**) and NDP-L-rhodinose (**13b**)

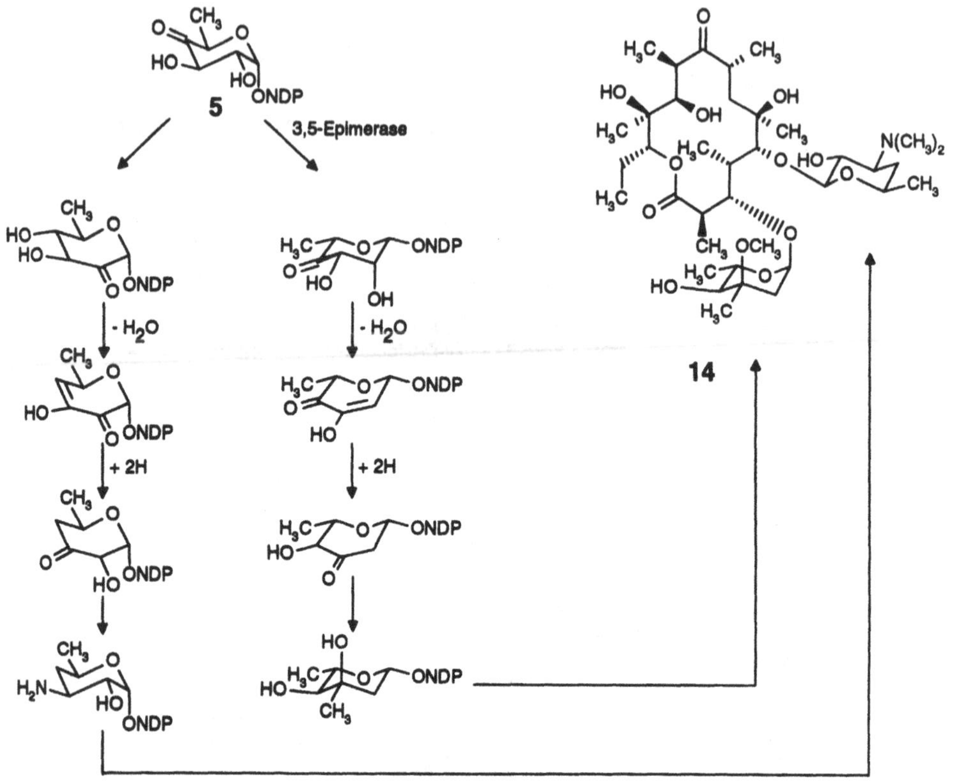

Scheme 7. Hypothetical deoxygenation pathways to the two deoxysugar moieties in erythromycin A (14)

mary, the 2-deoxygenation step in deoxysugar biosynthesis seems to proceed by different routes; however, mechanistic details remain to be elucidated.

2.1.4
4-Deoxygenation

4-Deoxysugars are found in a few antibiotics and other bioactive compounds, most notably in the antibiotics erythromycin A (14) [58, 24] and spectinomycin (15) [41] as well as in the α-glucosidase inhibitor acarbose (16, ring B), an important orally active natural antidiabetic agent [59, 60] (Scheme 8).

As already shown for the 3- and 2-deoxysugars, a NDP-4-keto-6-deoxyglucose 5 is assumed to be a precursor (Scheme 7) of D-desosamine and has been proven indirectly to be the precursor of 4,6-dideoxyhexose moiety of spectinomycin (15) [41], since stereochemical analysis showed that the configuration of the 6'-CH$_3$ group in 15 is inverted when fed with stereochemically labeled (6R)- and (6S)-[6-^{14}C, 4-^{2}H, 6-^{3}H]glucose. This inversion at C-6' points to the involvement of a dTDP-glucose oxidoreductase (see Sect. 2.1.1., Scheme 2). Since the 4,6-di-

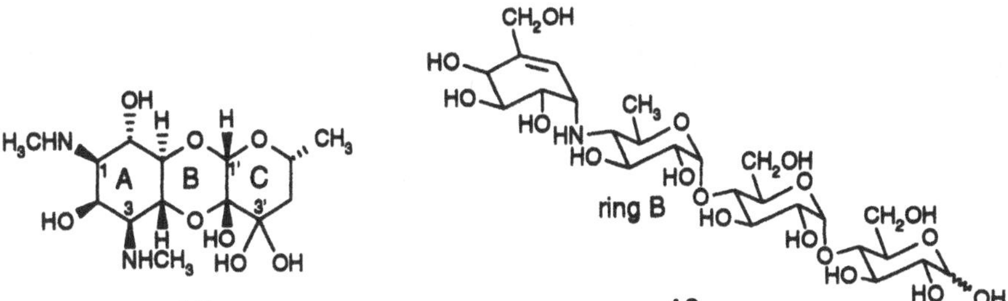

Scheme 8. Spectinomycin (15) [41] and the α-glucosidase inhibitor acarbose (16)

deoxysugar moiety (ring C) in spectinomycin carries a keto group at the 3-position (C-3', shown in 15 in the hydrate form), the 4-deoxygenation may be catalyzed by a pyridoxamine-linked deoxygenase in a manner similar to the 3-deoxygenation in ascarylose biosynthesis (see Sect. 2.1.2). However, comparison of the genes for deoxysugar biosynthesis with those for biosynthesis of NDP-ascarylose (7) revealed no evidence for a homolog of NDP-6-deoxy-$\Delta^{3,4}$-glucoseen reductase in the erythromycin producer *S. erythraea*. This enzyme is required for the C-3 deoxygenation mechanism found in the biosynthesis of 7 [53].

2.1.5
5-Deoxygenation

Since a 5-deoxygenation would preclude the formation of a pyranose ring, there are no "real" 5-deoxysugars. Noteworthy exceptions are the 5-amino-5-deoxysugars (piperidinoses), e.g. nojirimycin (17), 1-deoxynojirimycin (18), mannojirimycin (19) and galactostatin (20) [59, 61, 62] (Scheme 9). Typically, these "sugars" act as glycosidase inhibitors; 17 and 18 inhibit α-glucosidase, 19 α-mannosidase, and 20 β-galactosidase. Also, structural elements, such as the tetrahydropyrrol ring found in bulgecin A (21) [63, 64] or SQ 28504 [65, 66] may biosynthetically derive from glucose, and cyclopentanes, such as in pactamycin [25] and the allosamizoline unit 22 in the chitinase inhibitor, allosamidine (23) [67–69], have been proposed or proven, respectively, to originate from the sugar pool. Thus, they can be viewed as a 5-amino-2,3,5-trideoxyhexonic acid and as 1,5- (in pactamycin) or 2,5-dideoxysugars (22), respectively [25, 68, 59].

Biosynthetic studies on 1-deoxynojirimycin (18) using various organisms revealed that the "5-deoxygenation" indeed is a 2-deoxygenation, since D-glucose, the biosynthetic precursor of both 18 and 19 (mannojirimycin), undergoes a C-2/C-6 cyclization and subsequently is "inverted" during its processing to 19 and 18 [61, 62], i.e., label from C-1 of glucose ends up in C-6 of 18, and 19 was shown to be formed en route as the first aminosugar. The 2-deoxygenation of glucose occurs through transamination after isomerization to fructose (Scheme 10). The pathway outlined in Scheme 10 is supported by incorporation experiments with variously [^{2}H]- and [^{13}C]-labeled glucoses and by enzyme assays [61, 62].

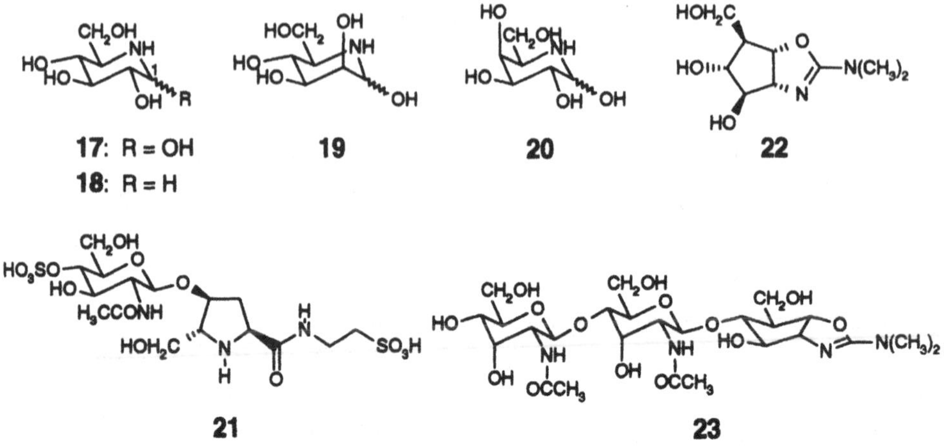

17: R = OH
18: R = H

19

20

22

21

23

Scheme 9. 5-Deoxysugars: The 5-amino-5-deoxysugars (piperidinoses) nojirimycin (**17**), 1-deoxynojirimycin (**18**), mannojirimycin (**19**) and galactostatin (**20**). The tetrahydropyrrol moiety of bulgecin A (**21**) and the allosamizoline unit **22** of allosamidine (**23** can also be considered as 5-deoxysugars. The derivation of **22** from the pool has already been proven

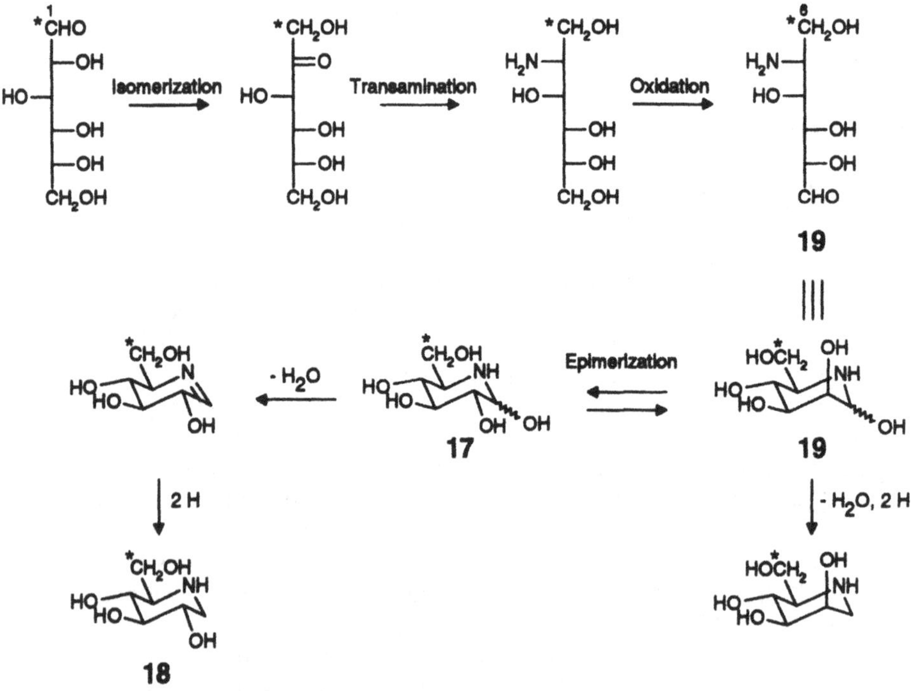

19

17

19

18

Scheme 10. Biosynthetic formation of 1-deoxynojirimycin (**18**) and mannojirimycin (**19**). Note the "inversion" of the label (the 1-label becomes a 6-label), i.e., the "5-deoxygenation" is really a 2-deoxygenation

Extensive biosynthetic studies were carried out on various aminoglycoside antibiotics, typified by streptomycin (**24**), neomycin C (**25**), or spectinomycin (**15**), already mentioned above. This group of antibiotics (Scheme 11) consists entirely of moieties derived from carbohydrate metabolism, i.e., the molecules can be more or less characterized as "oligosaccharides" [41, 70–72] (for further information, see also Sect. 3). In this context, the biosynthetic formation of 2-deoxystreptamine (**27**) is most important. This aminocyclitol core moiety of neomycin (streptamine **26** is the analogous aminocyclitol moiety of streptomycin) is biosynthesized entirely from D-glucose, and, within this biosynthetic sequence (Scheme 12), a formal 5-deoxygenation step occurs through trans-amination of the 3-keto group in the intermediate 2-deoxy-*scyllo*-inosose (**28**). Note, that the 3-position arises from C-5 of glucose! A similar transamination step also leads to the actinamine moiety of **15** via myo-inositol [41]. The conclusions presented in the biosynthetic mechanism shown in Scheme 12 [71a]

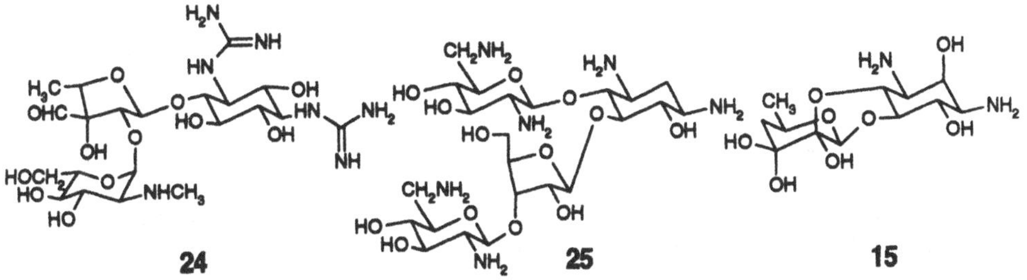

Scheme 11. The aminoglycoside antibiotics, typified by streptomycin (**24**), neomycin C (**25**), and spectinomycin (**15**)

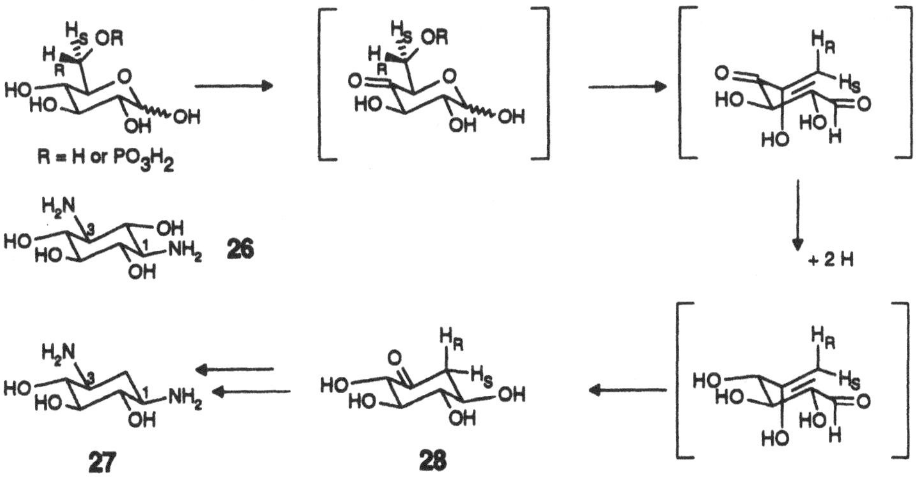

Scheme 12. 5-Deoxygenation during the biosynthesis of 2-deoxystreptamine (**27**; the core moiety of neomycin **25**, streptamine **26** is its analog in streptomycin **24**) through transamination of the intermediate 2-deoxy-*scyllo*-inosose (**28**). Note that C-5 of glucose becomes C-3 of **26** and **27**

are based on several incorporation experiments with growing cultures of *Streptomyces fradiae* using variously labeled glucose derivatives [71a, b] as well as on studies using cell-free preparations of the same organism [71c]. Interestingly, this cyclization reaction resembles the dehydroquinate synthase reaction of the shikimate pathway [2].

2.2
Glycosyl Transfer Steps and the Formation of Oligosaccharide Chains

Antitumor antibiotics, e.g., anthracyclines and angucyclines, are often equipped with single deoxysugar moieties, such as the aminosugar daunosamine in daunomycin (**29** = daunorubicin) and adriamycin (**30** = doxorubicin). However, some cytostatics are even characterized by one or more di- to hexasaccharide chains. Impressive examples for the latter group are shown in Scheme 13 and include the anthracyclines arugomycin (**31**) [73a] and viriplanin (**32**) [59, 73b, 73c], the angucyclines landomycin A (**12**) and vineomycin A_1 (**33** = P1894B [74]), or the aureolic acid antibiotics [75] mithramycin (**34**) [75a, 75b] and UCH9 [75c] (**35**). Also other bioactive compounds, e.g., the above mentioned amino glycoside antibiotics, the pseudotetrasaccharide acarbose (**16**) or orthosomycin antibiotics, such as avilamycin A [76] (**36**, excluding the small polyketide-derived aromatic ring), can, from their structures, roughly be viewed as "oligo-deoxysaccharides" (for further examples see also the chapter on angucyclines in this issue [77]). Although some details of the biosynthetic formation of the monosaccharide building blocks and the genes coding for the enzymes controlling these processes (see above and in Sect. 3) are already known, much less can be said about the exact biosynthetic stage at which the glycosyl transfer steps occur. Moreover, hardly any data exist on the assembly of oligodeoxysaccharide chains and the substrate specificity of the glycosyl transferases involved. For instance, how many glycosyl transferases are necessary for the construction of molecules like mithramycin (**34**) or landomycin A (**12**)? In general, it has been assumed that glycosyl transfer always occurs as the terminal step of a biosynthetic sequence, perhaps due to the ascribed lability of the glycosidic bond. This assumption has to be questioned, as briefly demonstrated below for selected examples of polyketide oligosaccharides (see Sect. 2.2.1), since in oligosaccharide chain formation, it is believed that these structural moieties arise linearly through stepwise glycosyl transfer of one (completely elaborated) sugar building block after the other.

2.2.1
Exact Stages of Glycosyl Transfer Steps in Biosynthetic Sequences of Polyketide Oligodeoxysaccharides

Although there are remarkable exceptions (see below), it has nonetheless often been shown that glycosyl transfer happens always at the very end of a biosynthetic sequence. For example, the biosynthesis of aclacinomycin A (**37**), a less cardiotoxic anthracycline than doxorubicin (**30**), proceeds via aklavinone (**38**)

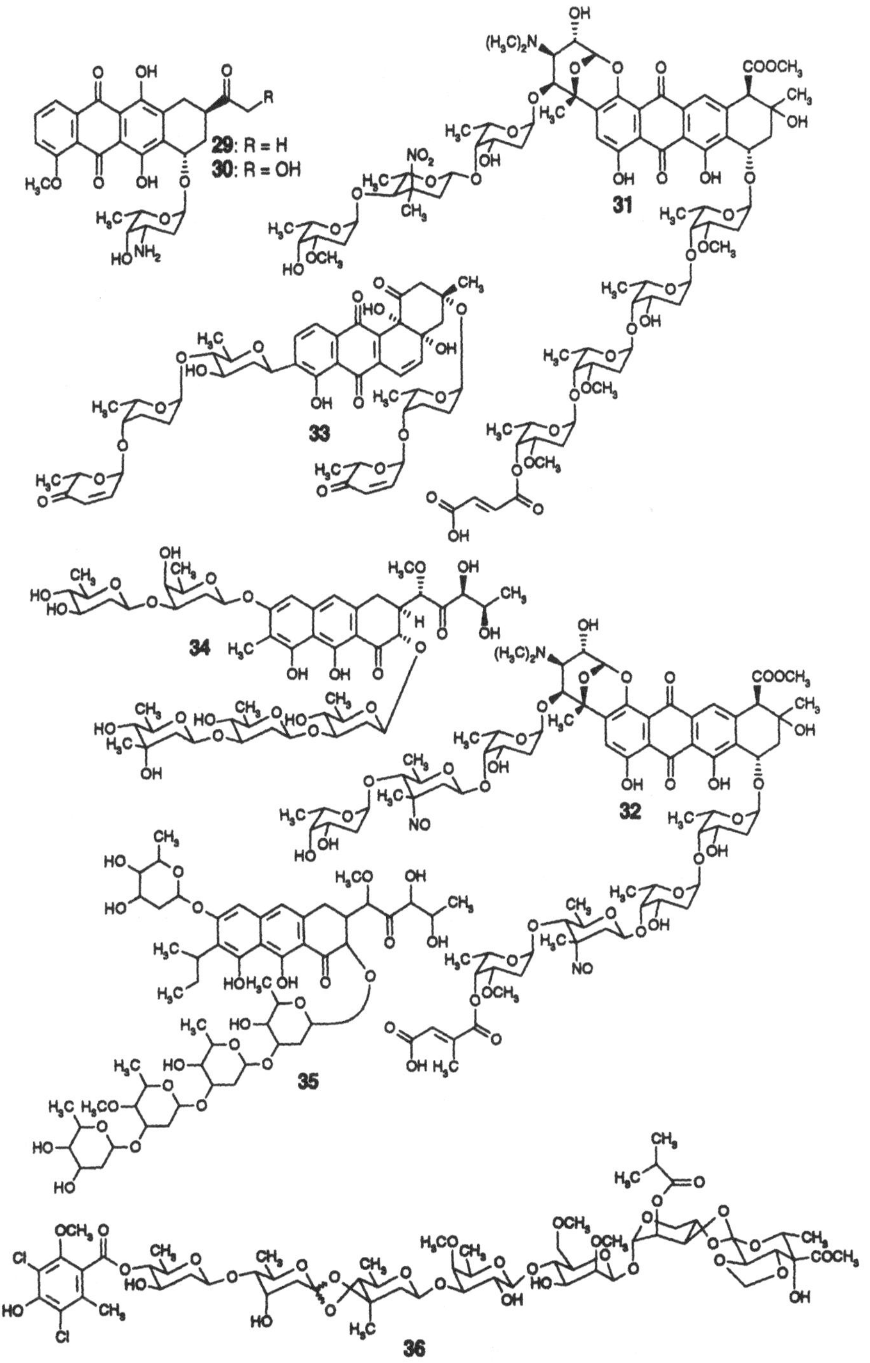

Scheme 13. The antitumor antibiotics daunomycin (**29** = daunorubicin) and adriamycin (**30** = doxorubicin) are polyketides with an attached deoxysugar moiety. Polyketide oligodeoxy-saccharides characterized by one or more deoxysaccharide chains include the anthracyclines arugomycin (**31**) [73a] and viriplanin (**32**), the angucyclines landomycin A (see **12** in Scheme 6) and vineomycin A₁ (**33** = P1894B [74]), and the aureolic acid antibiotics mithramycin (**34**) and UCH9 (**35**). Also, orthosomycin antibiotics, such as avilamycin A (**36**) can be considered as poly-ketide oligodeoxysaccharides

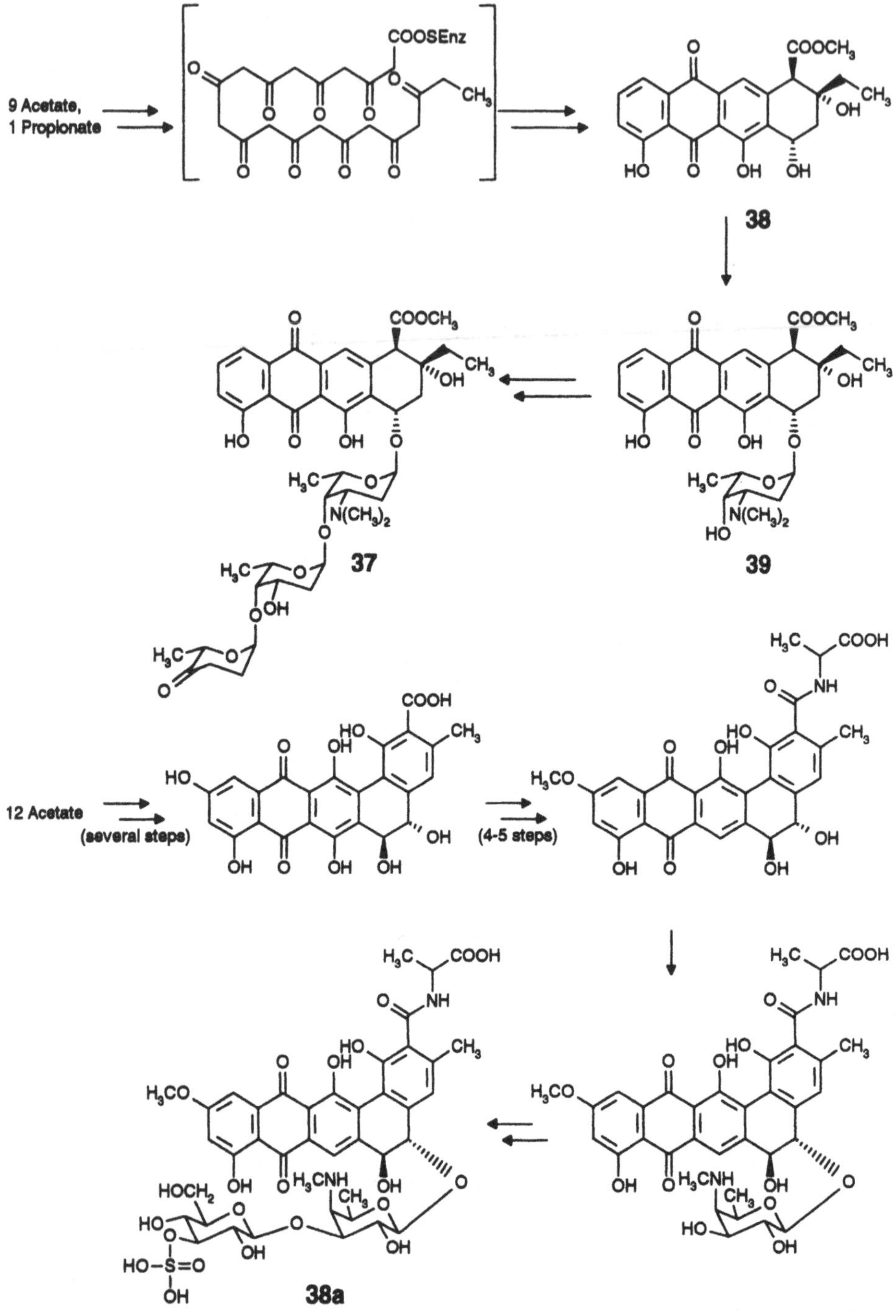

Scheme 14. The biosynthesis of aclacinomycin A (37 proceeds via aklavinone (38) and the monosaccharide aklavin (39), i.e., the glycosylation steps follow the complete construction of the aglycon. A similar process occurs in the biosynthesis of pradimicin S (38a); the glycosyl transfer steps occur after complete assembly of the polyketide aglycon moiety

and the monosaccharide aklavin (**39**), i.e., the glycosylation steps follow the complete construction of the aglycon (Scheme 14) [65].

Also, recent studies on pradimicin S (**38a**), an antifungal dihydrobenzo-[a]naphthacenequinone antibiotic with anti-HIV activity from *Actinomadura spinosa* AA0851, revealed a similar picture [78]. Bioconversion experiments and the use of blocked mutants and enzyme inhibitors showed that the glycosyl transfer steps occur after complete assembly of the polyketide aglycon moiety, including all post-polyketide oxygenation and reduction steps, and after the amino acid side chain and an additional methyl group were attached (Scheme 14, for further details see ref. [78]).

For landomycin A (**12**), it was suggested that glycosyl transfer occurs prior to biosynthetic completion of the aglycon moiety landomycinone (**40**), since the 8-OH group of the naphthazarine chromophore system of **40** does not favor sugar transfer, as was shown by model glycosylation experiments [79]. In earlier biosynthetic studies it was demonstrated that the oxygens at C-6 and C-7 derive from molecular oxygen, although they are located at positions which are derived from the acetate carboxyl group of [80–82]. Why are these two oxygen atoms eliminated from positions, in which they later have to be re-introduced, if not to facilitate the biosynthetic glycosyl transfer? But recent bioconversion studies on landomycin A (**12**) using [^{14}C]-labeled putative intermediates [81, 82] showed that landomycinone (**40**) is the ultimate substrate for the glycosyl transfer cascade to **12**, and the reasons for the elimination and re-introduction of oxygen at the 6- and 7-positions remain obscure (Scheme 15, for further details, see also below, Scheme 20a, and the angucycline chapter, Sect. 10 in [77]).

The C_2 symmetrical macrodiolide elaiophylin (**41**) consists of two identical polyketide chains which are linked to a deoxyfucose at C-13 [83]. Here, it was assumed that macrodilactone formation is the last biosynthetic step, and only one glycosyl transfer step is necessary prior to lactonization. However, incorporation experiments with the complete (glycosylated) octaketide half **42** failed. Later, after changing the fermentation parameters, the asymmetrical mono-glycosyl derivative **43** was isolated, indicating that the two glycosyl transfer steps take place after generation of the macrodiolide **44** [83–85] (Scheme 16).

Nevertheless, a few biosynthetic studies have shown that glycosyl transfer can occur much earlier, sometimes well before construction of the aglycon moiety is complete. In the perhaps best-known example, the biosynthesis of daunomycin (= daunorubicin, **29**) and adriamycin (= doxorubicin, **30**), glycosyl transfer of daunosamine is postulated to occur either at the aklavinone (**38**) stage [86], as in the biosynthesis of aclacinomycin A **37** (see above), or after 11-hydroxylation to ε-rhodomycinone [87], i.e., six to seven biosynthetic steps prior to completion of adriamycin formation (**30**, Scheme 17) [88].

It may be argued that this unusual biosynthetic sequence is due to the unique composition and organization of the *dps* (*dauno-* or *doxorubicin p*olyketide *s*ynthase) genes of *Streptomyces peucetius* [88]. But, also in more typical biosyntheses of polyketide oligodeoxysaccharides, for example, the macrolides erythromycin A (**14**) [24, 58, 89] in *Saccharopolyspora erythraea* and tylosin in *Streptomyces fradiae* [90], or urdamycin biosynthesis in *Streptomyces fradiae* Tü2717 [77, 91,

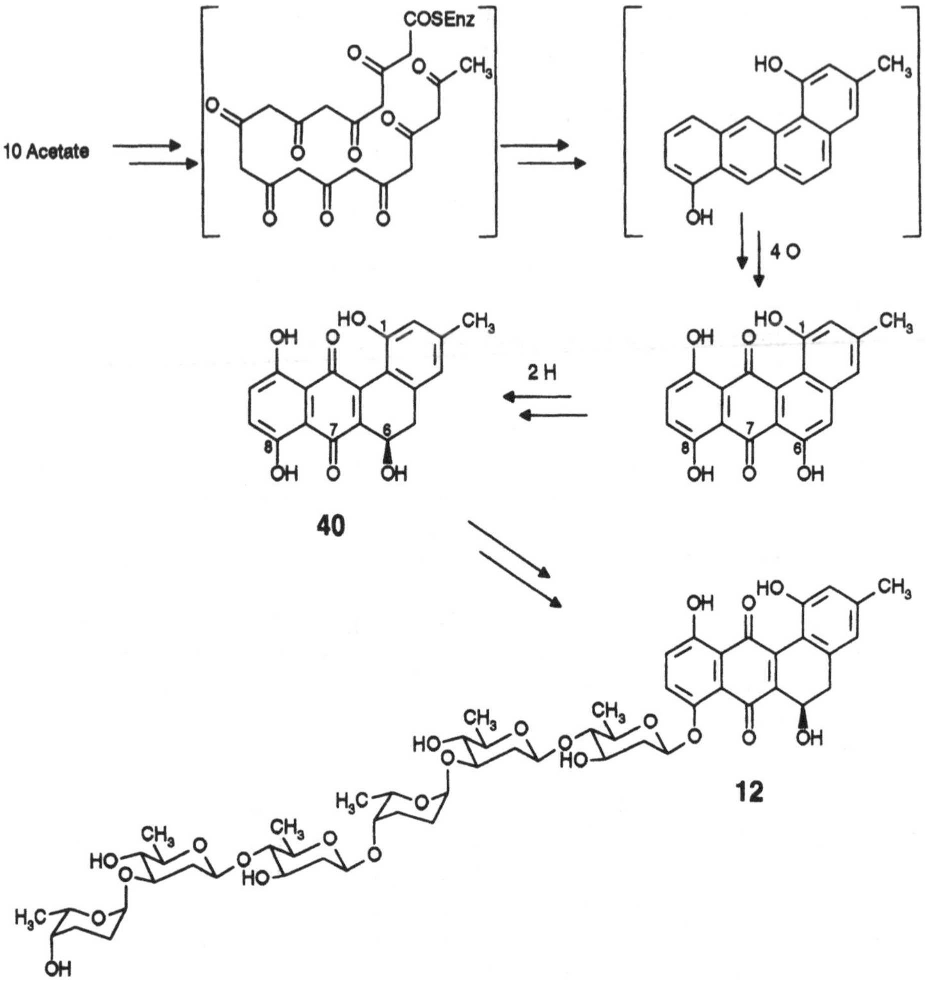

Scheme 15. The biosynthesis of landomycin A (**12**): landomycinone (**40**) is the ultimate substrate for the glycosyl transfer cascade to **12**

92], some glycosyl transfer steps by far precede the end of the biosynthetic sequence. In erythromycin A (**14**) biosynthesis (Scheme 18), the glycosylation step leading from erythronolide B (**45a**) to mycarosylerythronolide B (**45b**) and further to erythromycin D (**45c**) occurs prior to both C-12 oxygenation (C-12 hydroxylation) and the methylation of C-3″-O of the mycarosyl residue [24, 58, 89]. Similarly, for tylosin (**46**) biosynthesis (Scheme 18) it was found that the first sugar moiety (mycaminose) is already attached to tylactone (**47**) seven steps before the last O-methylation step from macrocin (**48**) to tylosin (**46**) [90].

The final product of the urdamycin biosynthetic sequence is urdamycin H (**49**) [93], which arises from decarbonylation of urdamycin C (**50**) [93, 94]. The latter is generated from tyrosine and urdamycin A (**51**) [95], a typical angu-

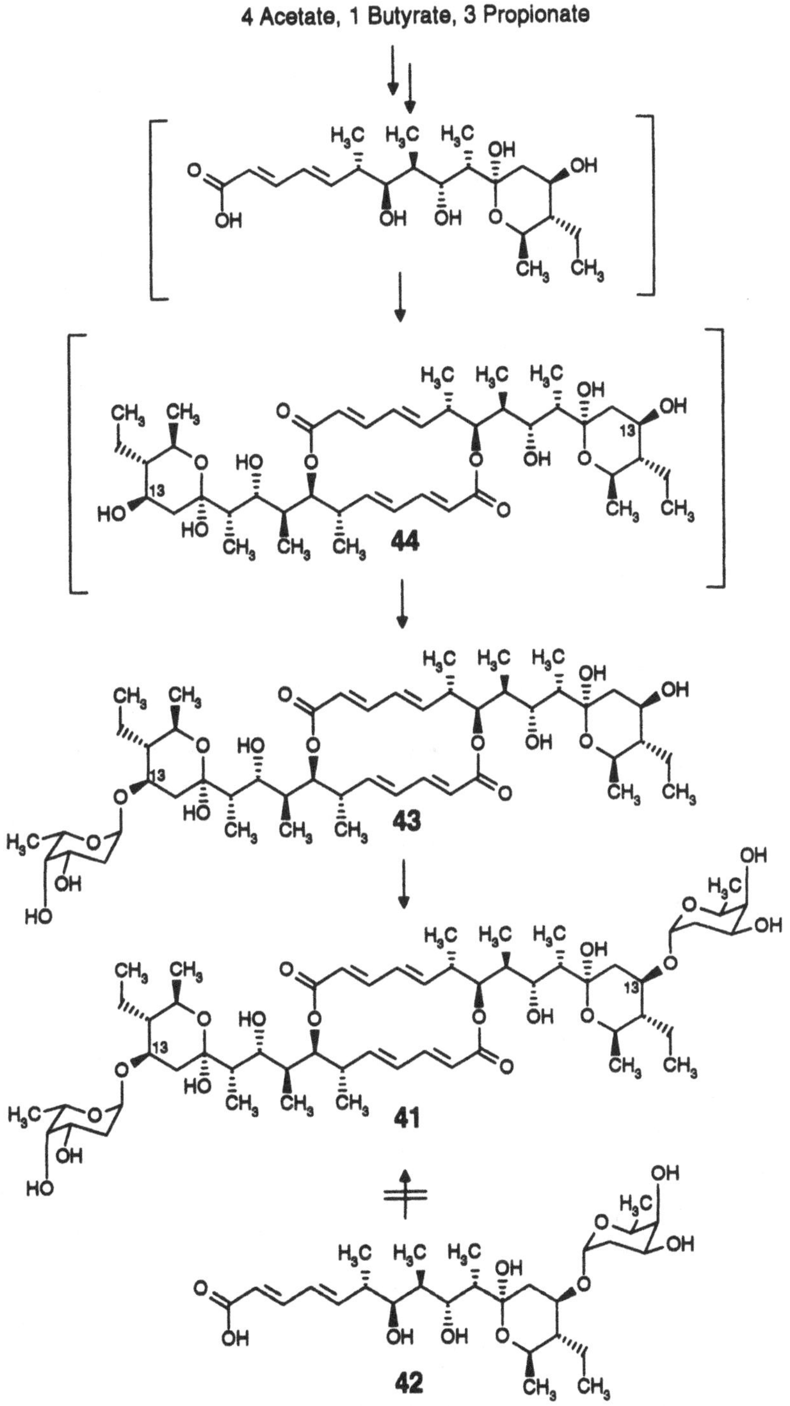
4 Acetate, 1 Butyrate, 3 Propionate
44
43
41
42

Scheme 16. Biosynthesis of the macrodiolide elaiophylin (**41**). The glycosylation steps occur at the very end of the biosynthesis

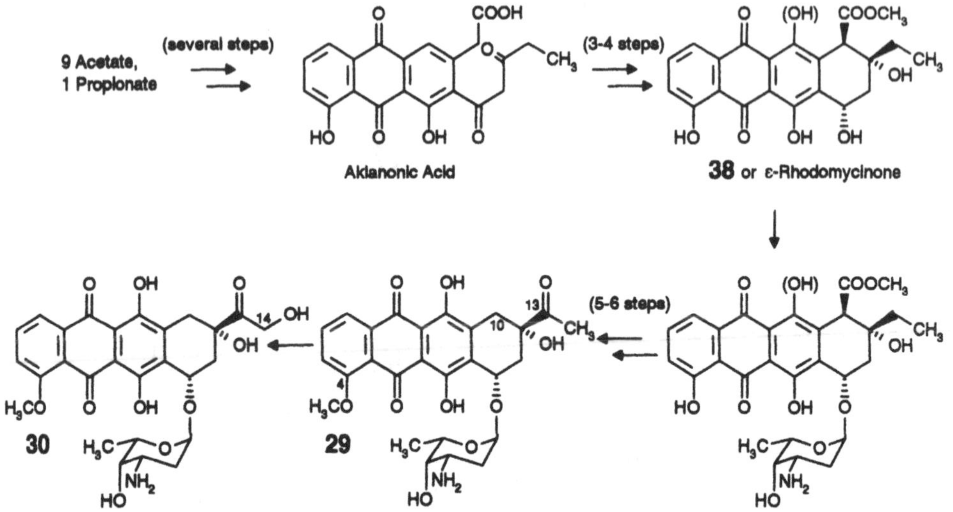

Scheme 17. The biosynthesis of adriamycin (30), demonstrating that glcosyl transfer can occur relatively early in the biosynthetic process

cycline-containing aquayamycin (52) as the aglycon moiety [77, 92]. The C-glycosylation step in the biosynthetic formation of aquayamycin (52) occurs prior to an oxygenation, a ketoreduction and a dehydration step through which the hydroxy group at C-12b and the 5,6-double bond are introduced [91]. Thus, considering the biosynthesis of the last metabolite of this sequence, urdamycin H (49), it is evident that the first glycosyl transfer occurs eight steps before the final ring contraction (50 to 49, Scheme 19; for more details see the chapter on angucyclines, Sect. 10) [77].

2.2.2
Biosynthesis of Oligodeoxysaccharide Chains

Acarbose (16, see also Sect. 2.1.4.), an α-glucosidase inhibitor and hence widely employed as an important oral antidiabetic, is considered a pseudotetrasaccharide. The so-called pseudo- or carbasugars, like the above mentioned aminocyclitol moieties of aminoglycoside antibiotics (see Sect. 2.1.5) or valienamine (53), entirely derive from the C_6 backbone of glucose or arise through gluconeogenesis. Thus, based on their biosynthesis, they have to be regarded as sugars. The pseudosugar moiety of acarbose (16), valienamine (53), as well as its presumed [25, 70] biosynthetic precursor valiolamine (54) are C_7-cyclitols with α-glucosidase inhibitory properties themselves; 53 is also found as an essential building block of the antibiotic validamycin A [25, 60, 70, 96, 97]. Biosynthetic studies on validamycin and acarbose by several groups of investigators [25, 60, 98] revealed that the C_7N backbone of valienamine arises through a [3+2+2] condensation from carbohydrate metabolism (Scheme 20); the proposed intermediate is a heptulose phosphate. Originally envisioned was a deri-

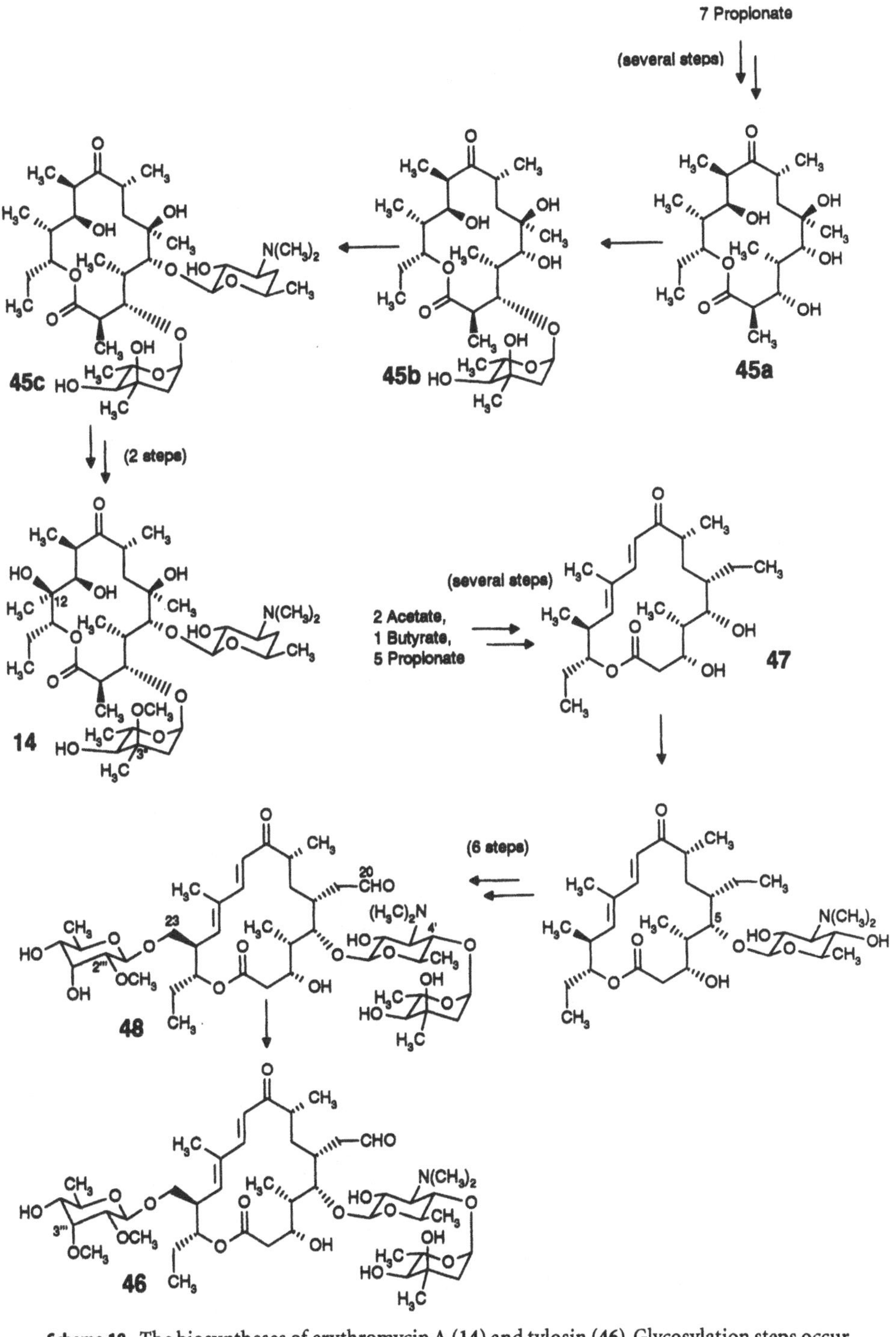

Scheme 18. The biosyntheses of erythromycin A (**14**) and tylosin (**46**). Glycosylation steps occur prior to the complete biosyntheses of the aglyca

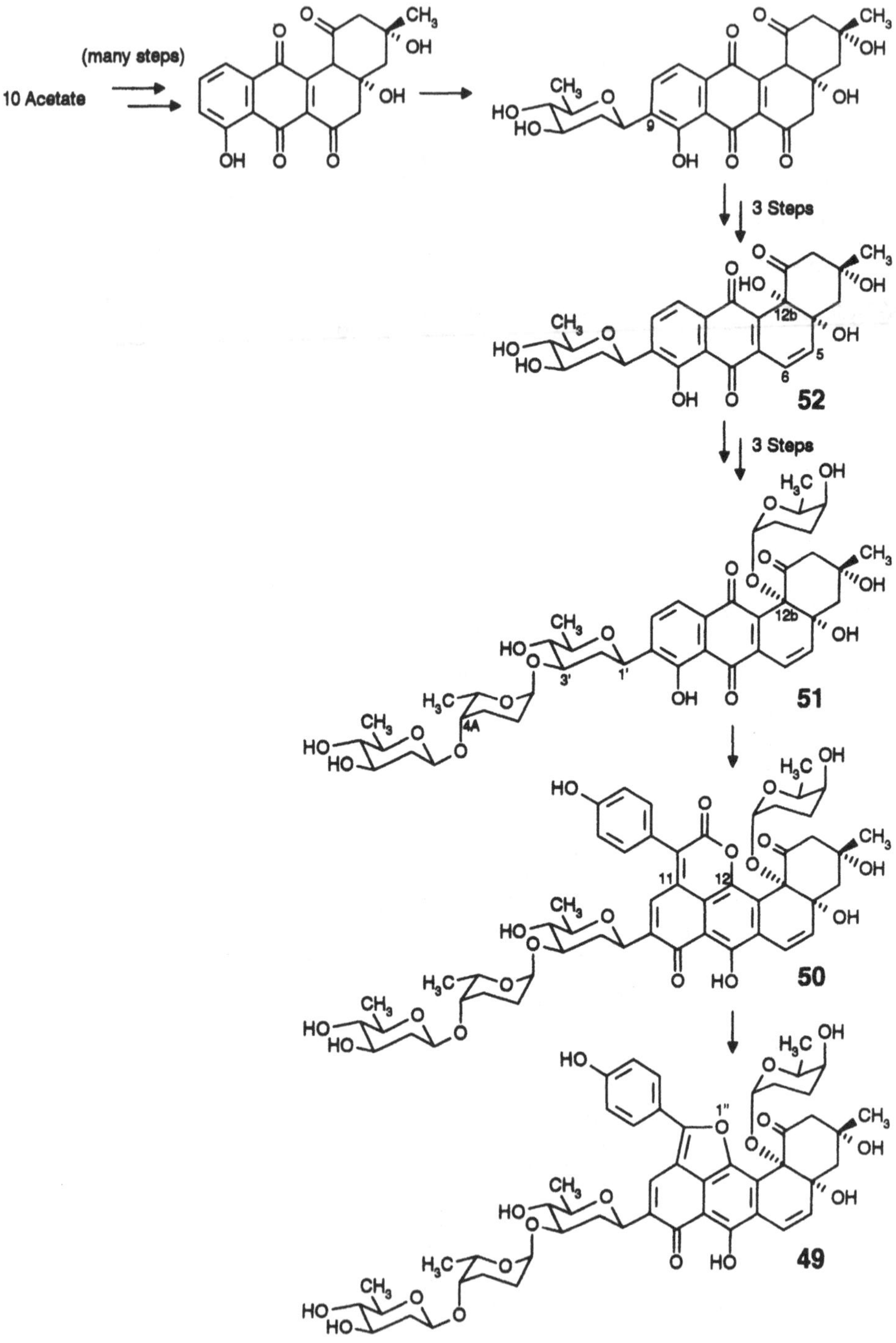

Scheme 19. The last steps of urdamycin biosynthesis: The first (C-) glycosylation step occurs long before the end of the biosynthetic cascade, with the final product urdamycin H (**49**)

vation of this C_7N building block from the shikimate pathway, a hypothesis which was refuted following biosynthetic studies [60].

The entire pseudotetrasaccharide acarbose (16) derives from this obscure C_7 sugar and three molecules of glucose [60]. There is evidence from incorporation experiments with labeled maltose and with derived from the fermentation medium [U-$^{13}C_3$]glycerol that the rings C and D are attached as an intact maltose unit, i.e., as a disaccharide. This is an example, in which a tetrasaccharide is assembled in a more convergent fashion than the generally postulated linear stepwise formation (Scheme 20a).

The hexasaccharide unit of landomycin A (12) is particularly fascinating, as it contains two repeating trisaccharide units, D-Oliv-1-4-D-Oliv-1-3-L-Rho (Oliv = olivose, Rho = rhodinose). In this case it was reasonable to assume that the glycan chain is biosynthetically constructed in a convergent manner in which the trisaccharide Oliv-Oliv-Rho was supposed to be formed prior to its linkage, first to landomycinone (40) and then to the trisaccharide intermediate landomycin E (55). Instead, a (more) linear assembly process (Scheme 16b) was found. This was deduced from bioconversion experiments on *Streptomyces cyanogenus* S-136 with [^{14}C]-labeled landomycins (labeled biosynthically from [1-^{14}C]acetate): The landomycin A congeners possessing a shorter sugar chain, landomycin D (56, two sugars) and landomycin B (57, five sugars; see Scheme 21), are indeed biosynthetic intermediates of landomycin A (12). Landomycin E (55) was recently isolated from a different organism, *Streptomyces globisporus* 1912 [99]. Although 55 has been postulated as an intermediate of 12 formation, it was never observed in the culture broth of *Streptomyces cyanogenus* S-136, even not in trace amounts [81, 82]. The molecule is extremly labile, cannot be produced in liquid medium, and therefore was not available for bioconversion studies. The fact that neither the mono- nor the tetraglycosyl analogs occur in S. *cyanogenus* S-136 [57, 100, 101] may be indicative of transfer of a dioliivoside during biosynthetic construction of the hexadeoxysaccharide chain. Alterna-

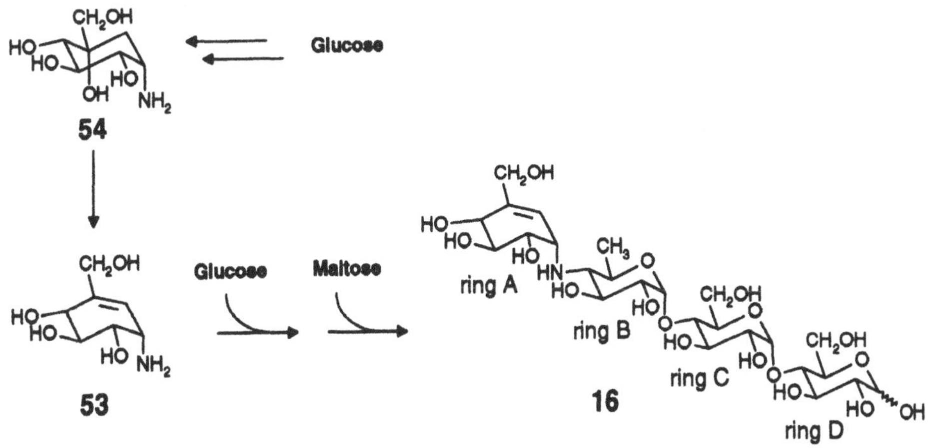

Scheme 20a, b. Disaccharide transfers are thought to be involved in the biosynthesis of acarbose (16, transfer of maltose) and landomycin A (12, transfer of dioliivoside units)

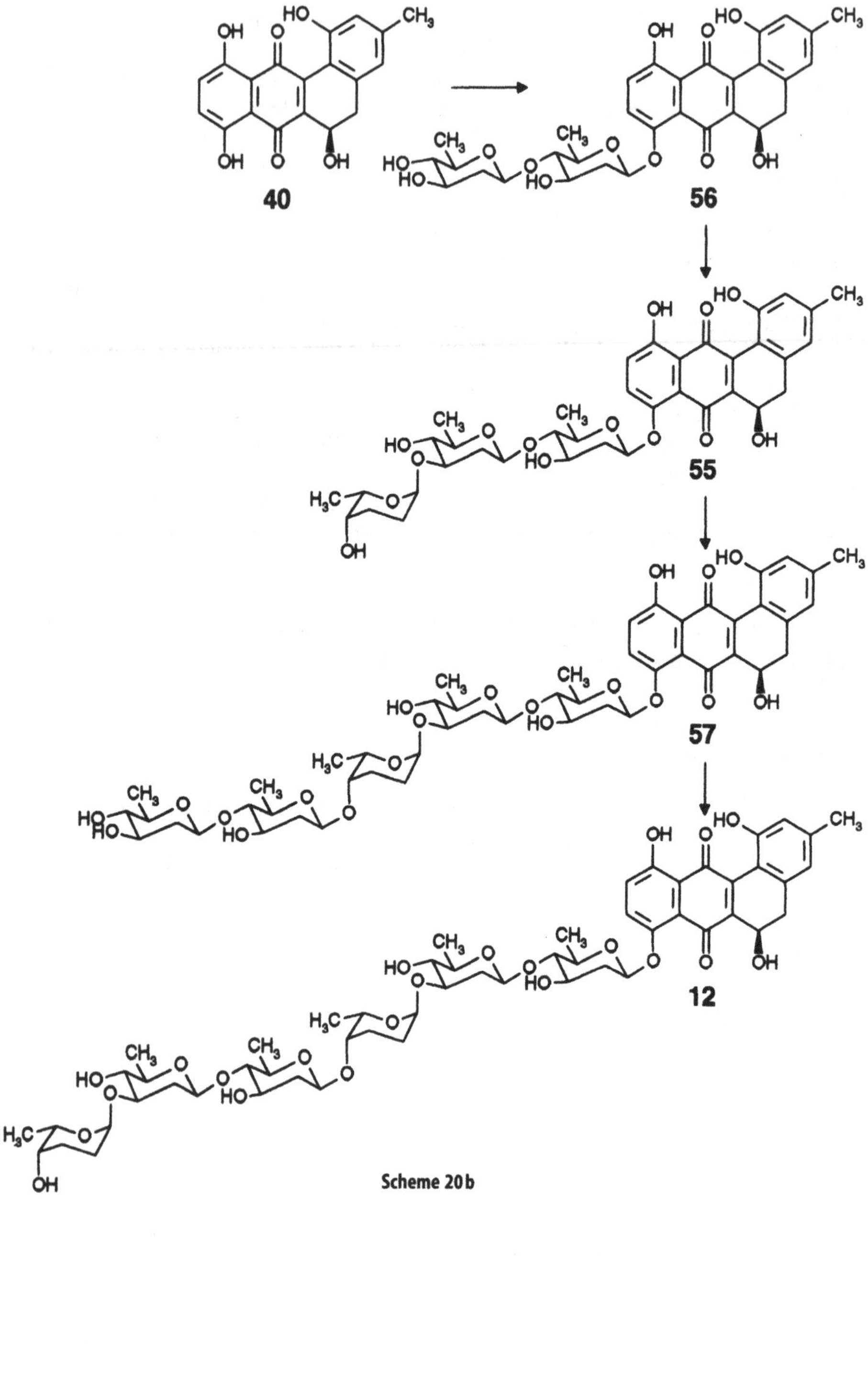

Scheme 20b

tively, this may be rationalized by kinetic reasons, i. e., the olivosyl transfer may proceed much faster than the rhodinosyl transfer (see Scheme 20b and also refer to Sect. 10 in the chapter on angucyclines).

From these examples it is clear that nature also provides convergent alternatives to the linear assembly of deoxysaccharides, as was, for example, observed in tylosin [90] or streptomycin (24) biosynthesis [72]. A linear strategy is certainly also true for other aminoglycosides, although in the case of neomycin biosynthesis, an alternative disaccharyl transfer has been discussed [70].

2.2.3
Specificity of Glycosyltransferases and Their Possible Use for Combinatorial Biosynthetic Approaches

Glycosyltransferases combine activated and already modified (deoxygenated, C-alkylated, etc.) sugar building blocks with an acceptor substrate, typically an alcohol, the so-called aglycon. The latter can be, of course, a sugar moiety itself. In many cases, these glycan units tremendously alter the structure and conformation of a molecule and often exert a major impact on the biological activity of the final natural product (see also Sect. 4). Only a few glycosyl transferases have been described thus far, most of them from higher organisms. They are responsible for particular steps in primary metabolism and are thus highly specific [102, 103]. With respect to the design of novel hybrid natural products, or for combinatorial biosynthetic approaches [104–109], the less specific glycosyl transferases from secondary metabolism would be more valuable. The recently analyzed mgt genes from Streptomyces lividans, code for a glycosyl transferase that inactivates macrolide antibiotics through glucosylation. It is quite substrate flexible and is known to be able to glucosylate 12-, 14-, 15- and 16-membered macrolides [110]. Another glycosyl transferase known from Streptomyces is encoded by the SnoT genes of S. nodosus [103]. In our own studies aimed at understanding the substrate flexibility of glycosyl transferases, parts of the elm gene cluster of the elloramycin (57) producer S. olivaceus Tü2353 (25 kb in the host plasmid 16F4) were transformed into the urdamycin (A: 51) producer Streptomyces fradiae Tü2717 [111]. The resulting transformant produced a hybrid antibiotic, 8–D-olivosyl-8-demethyltetracenomycin C (58), which clearly contains structural elements of both elloramycin (the aglycon) and urdamycin (the sugar moiety), indicating that the glycosyl transferase, either from S. fradiae Tü2717 or in cosmid 16F4, possesses a broad substrate flexibility (Scheme 21).

We favor the first alternative since further experiments with model substrates on S. fradiae revealed that presumably the same glycosyl transferase was able to C-glycosylate methylnaphthazarine (59). The resulting product, the olivosyl-naphthalinone 60, was also reduced by other enzymes of the urdamycin producer S. fradiae [81, 82]. The landomycin producer S. cyanogenus S-136, however, studied for reasons of comparison, was not able to glycosylate but only to reduce 59 to 61 (Scheme 22). This is a surprising result, since substrate 59 more closely resembles landomycin than the urdamycin chromophore; in fact, it actually matches the "western half" of landomycinone (40).

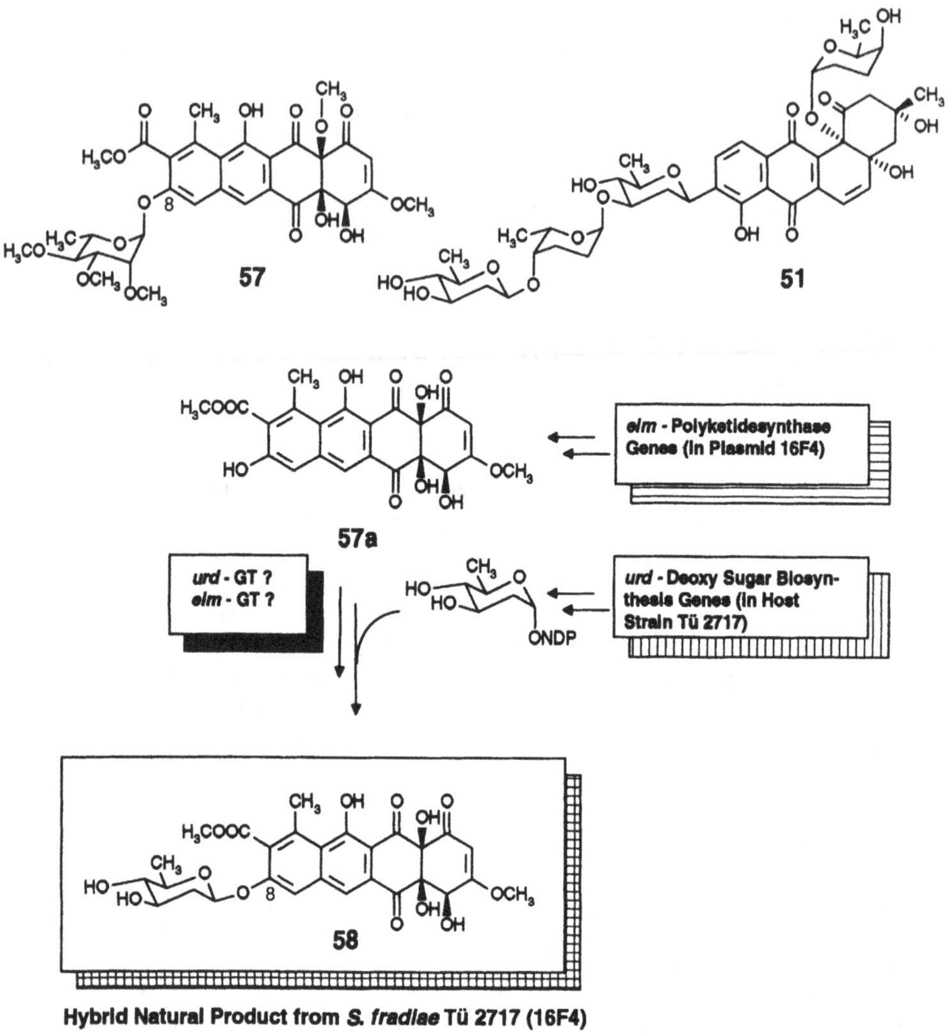

Scheme 21. The biosynthtic formation of the hybrid antibiotic 8-α-D-olivosyl-8-demethyl-tetracenomycin C (**58**) depends on genes from both parent organisms: The elloramycin (**57**) producer *Streptomyces olivaceus* Tü2353 provides the aglycon moiety (*elm* genes), the urdamycin (**51**) producer *Streptomyces fradiae* Tü2717 the deoxysugar moiety (*urd* genes). The origin of the combining glycosyltransferase (GT) is unclear

In other studies, the polyketide synthase genes of *S. fradiae* [112–114], responsible for the production of the aglycon moiety of the urdamycins, were deleted (*S. fradiae* Tü2717Δ7) [115], and this strain was used to host again the 16F4 genes of the elloramycin (**57**) producer. With this construct, *S. fradiae* Tü2717Δ7(16F4), olivosyltetracenomycin (**58**), originally a minor product (yield: ca. 8 mg l⁻¹), became the major product (yield: ca. 80 mg l⁻¹). When, for comparison, the *tcm* genes (pWHM1026) of the tetracenomycin producer *S. glaucescens* Tü49 were used instead of the 16F4 genes, there was absolutely no

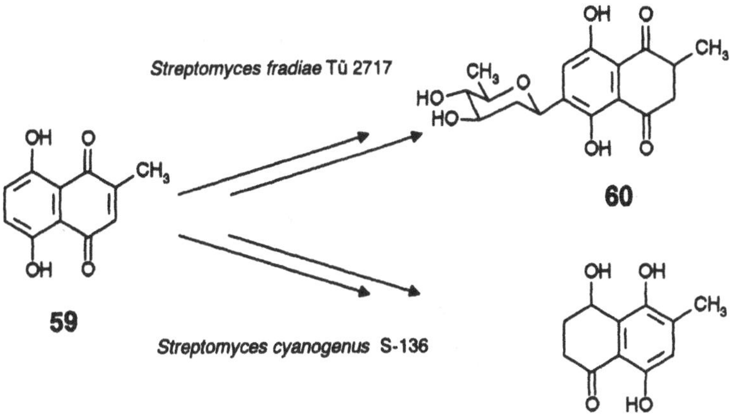

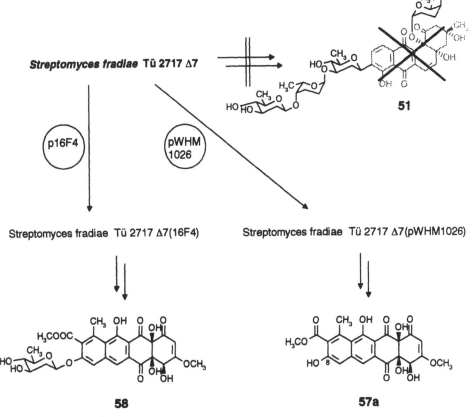

Scheme 22. *C*-glycosylation and reduction of methylnaphthazarine (59) by *Streptomyces fradiae* Tü2717 vs reduction only by *S. cyanogenus* S-136

Scheme 23. Biosynthesis of olivosyltetracenomycin (58) by the construction strain *S. fradiae* Tü2717Δ7(16F4). Substitution of the *elm* genes (cosmid 164F) by the *tcm* genes (plasmid pWHM1026) resulted in absolutely no production of 58, although the same aglycon, 8-*O*-demethyltetracenomycin C (57 a), is produced in high yields

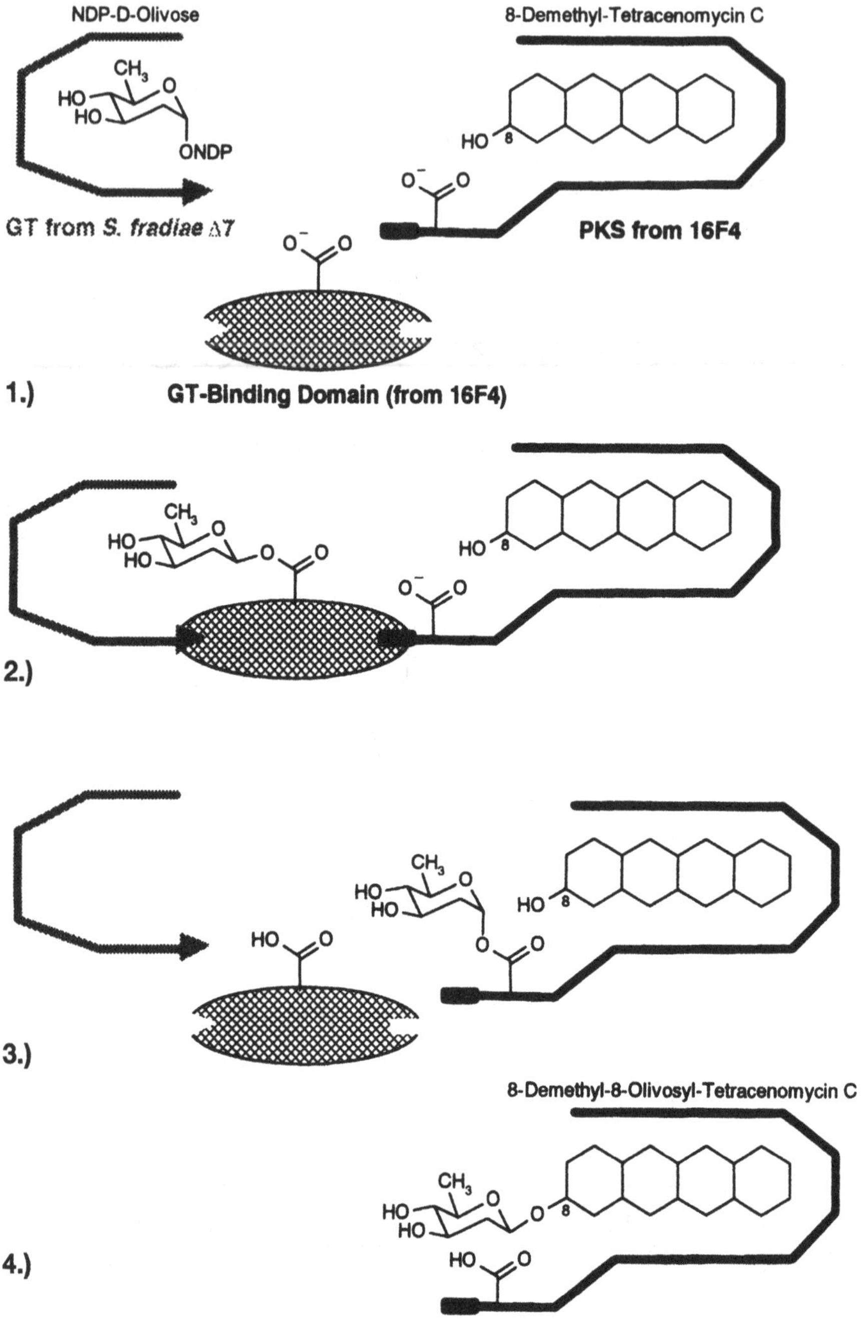

Scheme 24. Hypothesis regarding a glycosyl transfer supporting "domain" in cosmid 16F4

production of **58**. This is surprising, since 8-*O*-demethyltetracenomycin C (**57a**), the supposed substrate of the glycosyl transferase, is produced by cultures of this transformed strain, *S. fradiae* Tü2717Δ7 (pWHM1026), in high yield (Scheme 23). Our interpretation of this finding is that the *elm* genes, which are responsible for the production of **57a**, although similar, are nonetheless not identical to the *tcm* genes coding for the same aglycon. Since elloramycin bears a sugar moiety (permethyl-L-rhamnose) at the 8-position, it can be assumed that the *elm* genes responsible for the formation of **57a** also code for association of the glycosyltransferase/aglycon complex necessary for formation of **57** and the hybrid antibiotic **58**. This implies that the aglycon has to be "prepared" for a glycosyl transfer during its biosynthesis (Schemes 23, 24) [115b]. If this "preparing" domain could be found and altered through its encoding genes, this would be a major step in the design of hybrid glycosides in general. Alternatively, a glycosyl transferase in cosmid 16F4 that is highly flexible regarding the sugar could be considered.

These promising examples show that, at least some, if not considerable substrate flexibility exists for the alcohol (aglycon) moieties. Broad substrate specificity, which is a prerequisite for using glycosyl transferases in combinatorial biosynthetic approaches, can be tested quite easily in vivo using organisms that provide glycosyl transferases associated with certain, i.e., not derived from primary metabolism, deoxysugars such as olivose, rhodinose, and daunosamine. In such cases, it is not necessary [102] to make glycosyl transferase libraries and series of complementary deoxysugar nucleotide diphosphats for substrate specificity tests. In this context, it would be even better if glycosyl transferases that were flexible in their complementary activated sugar could be found or tailored.

2.3
Miscellaneous

So far, Sect. 2 of this article has generally dealt with biosynthetic studies on deoxysugars and deoxysugar-containing natural products. This last subsection provides a brief excursion into two aspects of carbohydrate metabolism that are important for biosynthetic studies of deoxysugars: (1) the pentose phosphate cycle in the metabolism of *Streptomycetes* and other microorganisms and (2) the "short activation pathway" of deoxygenated carbohydrates.

It is occasionally stated that studies with radioactive or stable isotope-labeled substrates give only partial or even misleading information [103] and therefore may no longer be regarded as state-of-the-art research methodology. However, progress in the complex field of nature's pathways to complicated molecules can only be achieved by a multidisciplinary approach, and significant advances are most certainly to be expected from biosynthetic studies using labeled compounds. Important aspects of deoxysugar metabolism, such as deoxygenations and glycosyl transfer, can be understood only from genetic and enzymatic studies that are combined with incorporation studies using sophisticated syntheses of multilabeled precursors and analyses of their "biosynthetic fate" in organisms or enzymes through chemical degradation and/or NMR studies of the resulting products.

2.3.1
Unexpected Results from Incorporation Experiments
("Scrambling" Through the Pentose Phosphate Cycle)

In several biosynthetic studies on deoxysugars, using either [^{13}C]-, [^{14}C]- or [^{2}H]-labeled glucose or glucose derivatives (e.g., recent studies on 1-deoxynojirimycin (18) [62, 61] and allosamidin (23) [68]) or [U-^{13}C$_3$]glycerol, (e.g., the studies on acarbose (16) [60] and urdamycin A (51) [116]), unexpected and intriguing incorporation results have been obtained. For instance, a "randomization" of [1-^{13}C]glucose by *S. subrubtilus* and *B. subtilis* was observed along with an unequal labeling of C-1 and C-6 of 18 [62]. In a related investigation an unexpectedly large loss of tritium was detected in the carbohydrate moieties of 23 when experiments with *Streptomyces sp.* were conducted using mixtures of [6-^{3}H]- and [1-^{14}C]-labeled glucosamine [68]. Furthermore, surprising observations were described [60b, 116], in that, after feeding of [U-^{13}C$_3$]glycerol, carbons 1, 2 and 3 (the "top halves") of glucose-derived deoxysugar moieties were less enriched than carbons 4, 5 and 6 (the "bottom halves"). This was rationalized by scrambling or by a rapid turnover in sugar metabolism [68]. Alternatively, it was suggested [60, 116] that larger amounts of the label of [U-^{13}C$_3$]-labeled glycerol recycle via glycerol phosphate into phosphoglyceraldehyde than into dihydroxyacetone phosphate. These two central C$_3$ building blocks of gluconeogenesis were postulated to not be in equilibrium, as they usually are (via triosephosphate isomerase). In general, glycolysis and gluconeogenesis are always been assumed to be the dominant metabolic pathways of microbial carbohydrate metabolism.

These unexpected and intriguing results, obtained from the feeding experiments, may be understood if one takes the pentose phosphate cycle into consideration (Scheme 25). This may well be the dominant carbohydrate metabolic pathway in *Streptomycetes* and other organisms during the production phase of secondary metabolites [25, 103], especially for the de novo formation of glucose (instead of gluconeogenesis). This was first observed by Rinehart et al. as part of their studies on neomycin and validamycin biosynthesis [25, 98, 117]. In this route, [U-^{13}C$_3$]glycerol only enters carbons 4, 5 and 6 of glucose 6-phosphate, i.e., they end up in the bottom half, if the sugar is drawn in the common Fischer projection (see Scheme 25). The unequal distribution of the [1-^{13}C]label in the biosynthesis of deoxynojirimycin (18) may be due to the breakdown of glucose to the triose phosphate pool via glycolysis, followed by partial channeling of this [3-^{13}C]-labeled C$_3$ building block into the pentose phosphate cycle, through which glucose, now [6-^{13}C]-labeled, is "reconstructed." In the cases observed, (more) directly [1-^{13}C]-labeled glucose competes with (less) [6-^{13}C]-labeled glucose formed through the breakdown/reconstruction detour. The loss of the [6-^{3}H]-label in the glucosamine moieties of allosamidin (23) may be explained by an "inversion" of glucose, which may proceed in a way similar to that found in the biosynthesis of deoxynojirimycin 18 (see Scheme 10) [62]. This is followed by partial dilution of so-formed [1-^{3}H]glucose through decarboxylation of C-1 via the pentose phosphate cycle. In contrast, the [1-^{14}C] label of glucose is "converted" into a [6-^{14}C] label and thus remains in the molecule. However it must be admitted that the explanation given here for the latter example is quite speculative.

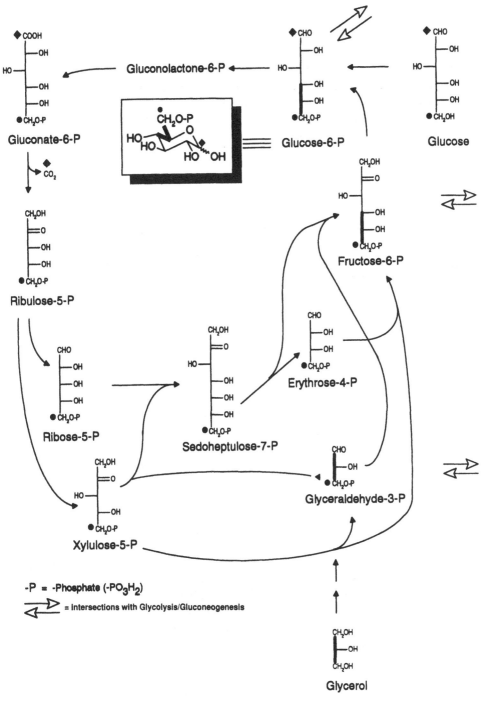

Scheme 25. ^{13}C labels in the pentose phosphate cycle

2.3.2
Short Activation of Deoxyhexoses

For the biosynthetic formation of deoxyhexoses, it is generally accepted that the deoxygenation steps and various other transformations occur on NDP intermediates, which are the activated substrates for glycosyl transfer. The activation pathway always starts with a hexose-6-phosphate which is then rearranged into the corresponding hexose-1-phosphate by a mutase, and is transformed into the activated NDP-sugar by a pyrophosphorylase, prior to all deoxygenation steps (see Scheme 6b for glucose). Since diphosphates are chemically labile, the risk is considerable that some of the activated sugar may be "deactivated" through hydrolysis anywhere during its long pathway to a tri-, tetradeoxy- or otherwise-modified sugar. These sugars would be lost for the final glycosyl transfer reactions if nature were not able to provide a possibility of "recycling," best achieved via a "short activation" pathway. This was first recognized and shown by Bekesi et al. [118] for L-fucose, a 6-deoxysugar which normally arises as an activated guanosyl diphosphate (GDP) derivative from GDP-D-mannose (see Sect. 2.1.1). By contrast, when the short activation pathway is operative, L-fucose is first converted into L-fucose-1-phosphate by a kinase and then further activated into GDP-L-fucose through a pyrophosphorylase-mediated coupling reaction (Scheme 26).

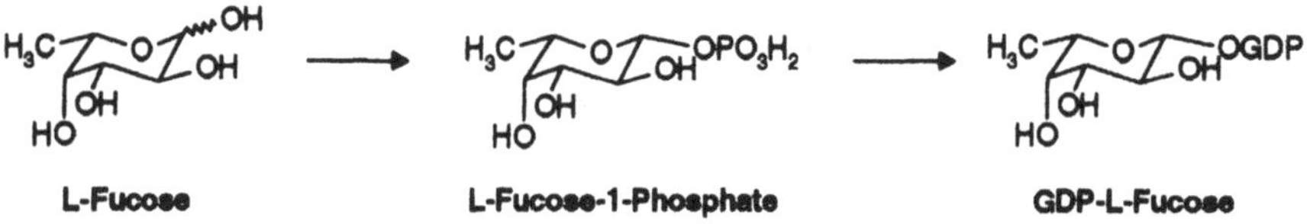

Scheme 26. "Short activation" of L-fucose

Additional evidence for an alternative pathway for the production of L-fucose came from experiments in mouse lymphoma cell lines blocked in the conversion of GDP-D-[^{3}H]-mannose to GDP-L-[^{3}H]-fucose. Reitman et al. observed that the block could be bypassed by growing the cells in the presence of L-fucose [119]. That this short activation pathway may be a general salvage mechanism is also indicated by the fact that 6-deoxy-D-glucose derivatives could be incorporated into neomycin [71]. The existence of a short activation pathway was recently proven in our laboratories [120] by successful incorporation of L-[2, 3-^{2}H$_2$]-rhodinose [121] into landomycin A (**12**). In this experiment, ^{2}H NMR spectroscopy of the product showed, that only the four protons of the two L-rhodinose moieties were labeled. In a control experiment, D-[2, 2, 3, 3-^{2}H$_4$]-rhodinose [121] was fed and, as expected, no incorporation into **12** was observed [120] (Scheme 27).

These results pave the way for more convenient biosynthetic investigations into deoxysugar biosynthesis, since it is now possible to perform incorporation experiments with isotope-labeled free deoxysugars whose NDP derivatives are postulated intermediates of certain deoxygenation pathways. For example, labeled L-aculose (**62**) is an important candidate compound that can be fed to

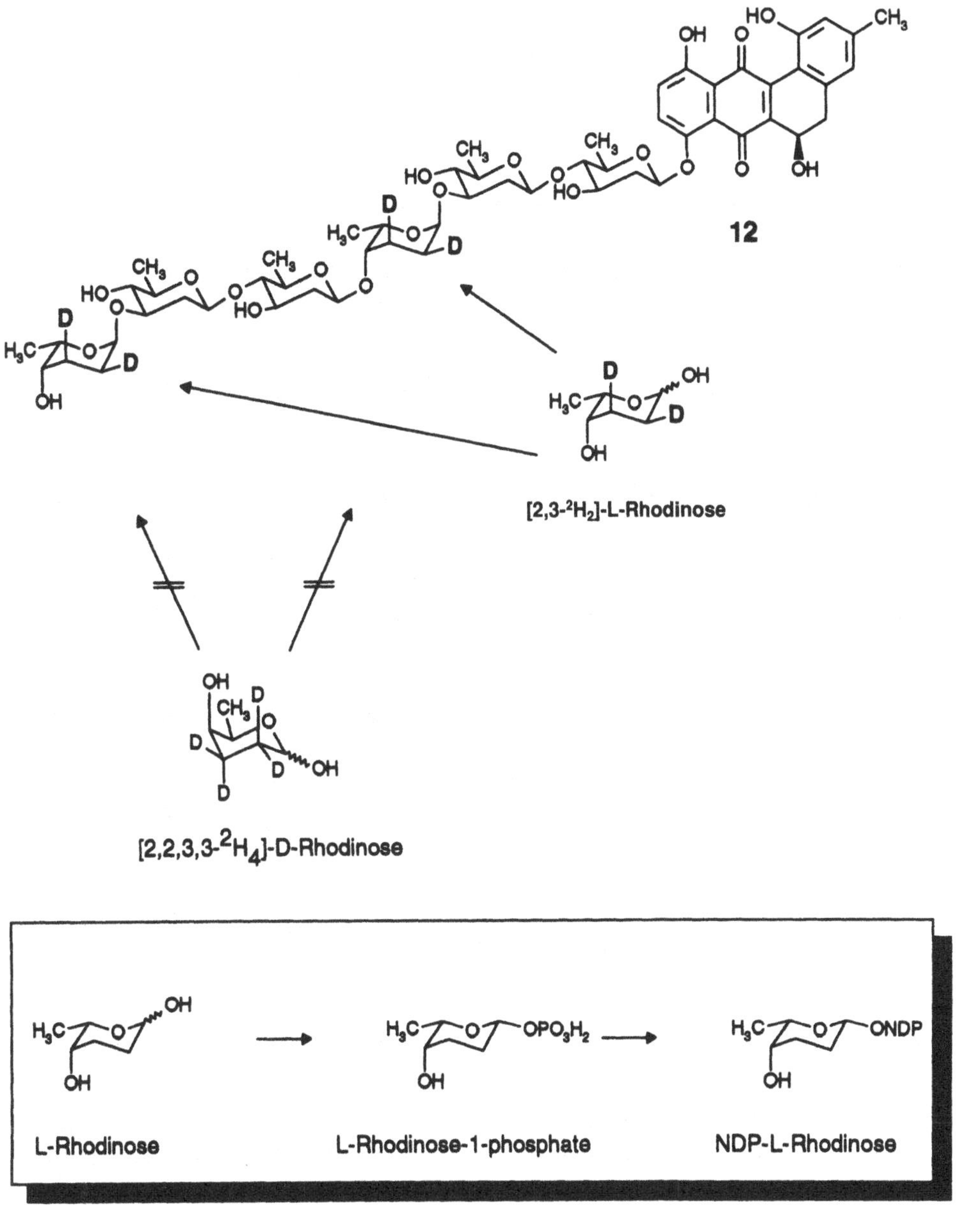

[2,3-²H₂]-L-Rhodinose

12

[2,2,3,3-²H₄]-D-Rhodinose

L-Rhodinose   L-Rhodinose-1-phosphate   NDP-L-Rhodinose

Scheme 27. Successful incorporation of [2, 3-²H₂]-L-rhodinose [121] into landomycin A (12) provides evidence for a short activation pathway

cultures of the landomycin producer *S. cyanogenus* S-136 to gain further insight into the formation of L-rhodinose (see Scheme 6, Sect. 2.1.1). A number of other opportunities arise from these findings. For example, it will be profitable to look for the activating genes involved in the short activation pathways, since the corresponding enzymes may exhibit a broad substrate flexibility and thus may be able to activate various deoxyhexoses, presumably at least all intermediates of a given deoxygenation cascade. In the future, such enzymes could serve as powerful tools in combinatorial biosynthetic (pathway engineering) approaches [104–109].

3
Molecular Biology and Enzymatic Aspects of Deoxysugar Biosynthesis

In the era of gene technology, molecular biological methods support biosynthetic investigations in a very constructive way, since elucidation of the corresponding biosynthetic genes ideally complements incorporation experiments. For example, the existence of a specific biosynthetic step can be assumed if an ORF (open reading frame) or an ORF domain of the corresponding enzymatic activity can be recognized among the biosynthetic genes of an organism. This is often possible, since DNA and protein sequence data regarding biosynthetic enzymes/activities can be obtained from gene data bases. Another interesting approach is to selectively disrupt or delete certain genes and then look for the products that accumulate in the blocked mutant. Finally, it is already possible to a certain extent to modify or even reconstruct biosynthetic genes in order to design novel "unnatural" or hybrid natural products, a process that will become even easier as more biosynthetic pathways are elucidated in detail. This will allow all enzymes involved to be detected indirectly via their corresponding genes; the latter can then be sequenced and assigned unambiguously to their respective biosynthetic steps. In addition, discovery of the genes required for deoxysugar biosynthesis will ultimately facilitate production of sufficient amounts of the corresponding enzymes (glycosyltransferases, glycosylases, pyrophosphorylases, etc.), thereby enlarging the potential of enzyme-based oligosaccharide syntheses (see Sect. 5.2).

As discussed in Sect. 2, current understanding of deoxysugar biosynthesis and the assembly of deoxysugar-containing oligosaccharide moieties is very limited. This may, in part, be due to the fact that deoxysugar formation takes place at the sugar nucleotide level. These sugar nucleotides are very difficult to isolate, because they do not accumulate under fermentation conditions and in many cases they are quite unstable. Due to this lack of potential substrates it has been very difficult to isolate proteins involved in the biosynthesis of deoxysugars or their assembly.

Molecular cloning has allowed isolation of biosynthetic gene clusters required for the synthesis of many antibiotics from Actinomycetes as well as gene clusters encoding antigen-specific polysaccharide formation. The function of these genes was deduced by comparing the encoded amino acid sequences with those of proteins whose function had already been determined. Identification of potential substrate binding sites, cofactor binding sites and various other motifs in the deduced amino acid sequences provided information about the possible function of the as yet undiscovered biosynthetic protein. Nonetheless, in many cases the function of the genes remained unknown, because no homologies could be found between the deduced amino acid sequence and the sequences of known proteins.

In recent years vector systems have been developed which allow the transfer of genes between Actinomycetes producing various antibiotics. By using these vector systems it has been possible to move biosynthetic genes from one strain into another, resulting in the production of hybrid antibiotics (see Sect. 2.2.3). In addition, preliminary experiments to overexpress genes from Actinomycetes in *E. coli* for the production of enzymatically active proteins have been carried out successfully. It will not be very long until more or even all enzymes involved in deoxysugar formation are available following this approach. Once the functions of deoxysugar biosynthetic genes have been elucidated the corresponding proteins will be an important source for the industrial production of nucleotide-activated sugars (see Sect. 5.2).

Presented below is an overview of known genes and gene clusters involved in the biosynthesis of deoxysugars and deoxysugar-containing oligosaccharide moieties of bioactive natural products. The discussion is based on literature available up until mid-1996.

3.1
Genes and Enzymes Involved in the Biosynthesis of Deoxysugars

As decribed above, not many biosynthetic genes coding for proteins involved in the biosynthesis of deoxysugars have been identified via characterization of the corresponding gene products. In many cases gene function has been assigned by comparing the deduced protein sequence with sequences of known proteins. However, a few genes have been expressed in heterologous hosts. In these cases proteins were purified and biochemically characterized. In the following, the most important biosynthetic genes and proteins involved in deoxysugar formation of antibiotics or of antigenic cell wall lipopolysaccharides in gram-negative bacteria are described.

3.1.1
Genes and Proteins Involved in the Conversion of Hexose1-Phosphate to NDP-Hexoses

There is good evidence that genes coding for NDP-hexose synthetases (hexose-1-phosphate nucleotidyltransferases, NDP-hexose pyrophosphorylases) are involved in the formation of deoxysugars in various organisms. As was mentioned before (see Sect. 2.1; Scheme 6b), these enzymes catalyze the conversion of a hexose 1-phosphate to a NDP-hexose.

Based on protein similarity, all NDP-hexose pyrophosphorylase genes, such as *StrD, StrQ* [72], *GraD* [122], *RfbA, RfbF, AscA, RfbM* [123] and *GlgC*, are related to each other [102]. An alignment of *GlgC* (adenosine-diphosphate-(ADP)-glucose synthetase) from *E. coli*, *RfbF* (glucose 1-phosphate cytidyltransferase) from *Salmonella enterica LT2* and *StrD* (dTDP-glucose pyrophosphorylase) from *S. griseus N2-3-11* is presented in Fig. 1. Structural studies have been carried on an ADP-glucose synthetase which was isolated from *E. coli*. The NH_2-terminal region of this enzyme has been postulated to be the activator binding site, whereas a region located in the center of the enzyme is thought to be the substrate binding site (Fig. 2) [124–127].

Recently, a glucose 1-phosphate thymidyltransferase from *Salmonella enterica* (strain LT2) has been characterized following expression of the gene

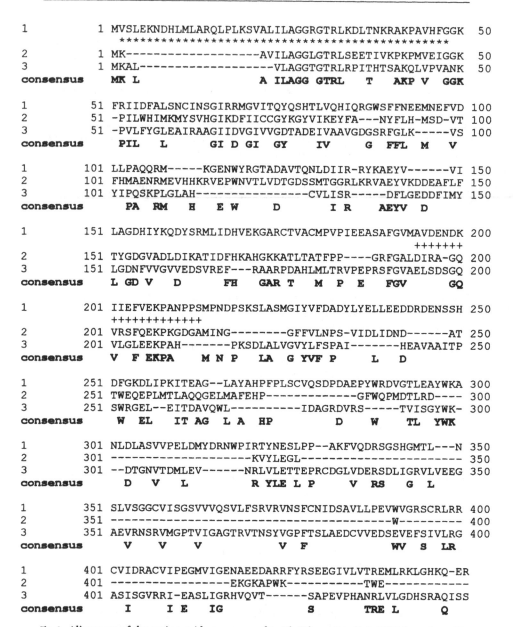

```
1      1 MVSLEKNDHLMLARQLPLKSVALILAGGRGTRLKDLTNKRAKPAVHFGGK  50
         **************************************************
2      1 MK-------------------AVILAGGLGTRLSEETIVKPKPMVEIGGK  50
3      1 MKAL-----------------VLAGGTGTRLRPITHTSAKQLVPVANK    50
consensus  MK L              A ILAGG GTRL     T   AKP V  GGK

1     51 FRIIDFALSNCINSGIRRMGVITQYQSHTLVQHIQRGWSFFNEEMNEFVD 100
2     51 -PILWHIMKMYSVHGIKDFIICCGYKGYVIKEYFA---NYFLH-MSD-VT 100
3     51 -PVLFYGLEAIRAAGIIDVGIVVGDTADEIVAAVGDGSRFGLK-----VS 100
consensus   PIL   L      GI D GI GY      IV     G  FFL  M   V

1    101 LLPAQQRM-----KGENWYRGTADAVTQNLDIIR-RYKAEYV------VI 150
2    101 FHMAENRMEVHHKRVEPWNVTLVDTGDSSMTGGRLKRVAEYVKDDEAFLF 150
3    101 YIPQSKPLGLAH----------------CVLISR-----DFLGEDDFIMY 150
consensus     PA  RM    H    E  W      D      I  R     AEYV    D

1    151 LAGDHIYKQDYSRMLIDHVEKGARCTVACMPVPIEEASAFGVMAVDENDK 200
                                                     ++++++
2    151 TYGDGVADLDIKATIDFHKAHGKKATLTATFPP----GRFGALDIRA-GQ 200
3    151 LGDNFVVGVVEDSVREF---RAARPDAHLMLTRVPEPRSFGVAELSDSGQ 200
consensus   L GD V    D       FH    GAR T    M   P   E   FGV       GQ

1    201 IIEFVEKPANPPSMPNDPSKSLASMGIYVFDADYLYELLEEDDRDENSSH 250
            +++++++++++++
2    201 VRSFQEKPKGDGAMING--------GFFVLNPS-VIDLIDND------AT 250
3    201 VLGLEEKPAH-------PKSDLALVGVYLFSPAI-------HEAVAAITP 250
consensus   V  F EKPA    M N P    LA  G YVF P       L      D

1    251 DFGKDLIPKITEAG--LAYAHPFPLSCVQSDPDAEPYWRDVGTLEAYWKA 300
2    251 TWEQEPLMTLAQQGELMAFEHP-------------GFWQPMDTLRD---- 300
3    251 SWRGEL--EITDAVQWL----------IDAGRDVRS-----TVISGYWK- 300
consensus   W   EL    IT AG  L A   HP           D    W    TL  YWK

1    301 NLDLASVVPELDMYDRNWPIRTYNESLPP--AKFVQDRSGSHGMTL---N 350
2    301 --------------------KVYLEGL----------------------- 350
3    301 --DTGNVTDMLEV------NRLVLETTEPRCDGLVDERSDLIGRVLVEEG 350
consensus     D  V    L       R YLE L P      V  RS    G  L

1    351 SLVSGGCVISGSVVVQSVLFSRVRVNSFCNIDSAVLLPEVWVGRSCRLRR 400
2    351 ------------------------------------------W------- 400
3    351 AEVRNSRVMGPTVIGAGTRVTNSYVGPFTSLAEDCVVEDSEVEFSIVLRG 400
consensus      V    V    V        V  F        WV  S  LR

1    401 CVIDRACVIPEGMVIGENAEEDARRFYRSEEGIVLVTREMLRKLGHKQ-ER
2    401 ----------------EKGKAPWK----------TWE-----------
3    401 ASISGVRRI-EASLIGRHVQVT------SAPEVPHANRLVLGDHSRAQISS
consensus     I    I E    IG        S       TRE L       Q
```

Fig. 1. Alignment of the amino acid sequences of *1: GlgC* from *E. coli*; *2: RfbF* from *S. typhimurium*; *3: StrD* from *S. griseus*. The amino acids present at least in 2 of 3 are defined as consensus. "*" indicates a postulated activator binding site and "+" indicates a postulated substrate binding site

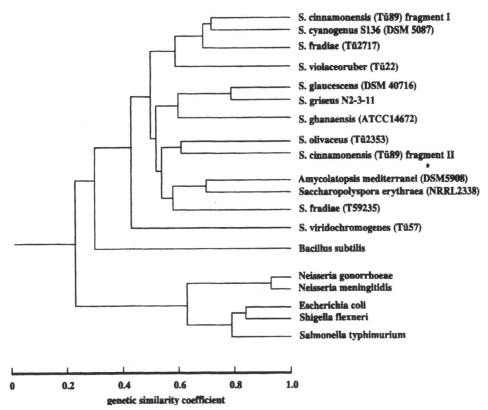

Fig. 2. Phylogenetic trees are based on genetic similarities. The sequences were taken from the following sources: *S. cinnamonensis* (Tü89) fragment 1 (EMBL X97834); *S. cyanogenus S136* (DSM 5087) (EMBL X97833); *S. fradiae* (Tü2717) (EMBL X97836); *S. violaceoruber* (Tü22) (gpL37334); *S. glaucescens* (DSM 40716) (EMBL X97837); *S. griseus* (N2-3-11) (sp P29782); *S. olivaceus* (Tü2353) (EMBL X97838); *S. cinnamonensis* (Tü89) fragment 2 (EMBL X97835); *Amycolatopsis mediterranei* (DSM 5098) (EMBL X97861); *Saccharopolyspora erythraea* (NRRL 2338) (gp L37354); *S. fradiae* (T59235) (gp U08223); *S. viridochromogenus* (Tü57) (EMBL X98039); *Bacillus subtilis* (sp P39630); *Neisseria gonorrhoeae* (sp P37761); *N. meningitidis* (gp L09188); *E. coli* (gp U23775); *Shigella flexneri* (sp P37777); *S. typhimurium* (sp P26391)

rfbA in *E. coli*. Kinetic measurements indicate that the enzyme acts via a ping-pong mechanism. Several substrates such as deoxythymidine triphosphate (dTTP), uridine triphosphate (UTP), α-D-glucose 1-phosphate and α-D-glucosamine 1-phosphate and α-D-xylose 1-phosphate were accepted, indicating a broad substrate specificity of this enzyme [128] (see also Sect. 5.2.1.2).

3.1.2
Genes and Proteins Involved in the Conversion of NDP-Hexose to NDP-4-keto-6-Deoxyhexose

The transformation of NDP-hexoses to NDP-4-keto-6-deoxyhexoses is catalyzed by NDP-hexose 4,6-dehydratases (see Scheme 2, Sect. 2.1.1). Genes coding for these dehydratases are highly conserved in Actinomycetes. DNA probes derived from *strE* from *S. griseus N2-3-11* have been utilized to detect chromosomal DNA fragments that presumably contain genes encoding enzymes responsible for the formation of 6-deoxysugars in various Actinomycete strains [129]. In addition, a polymerase chain reaction (PCR) method has been developed which can be used for the rapid amplification of DNA fragments containing 4,6-dehydratase genes from a wide range of Actinomycete strains. This approach allowed construction of a phylogenetic tree for the proteins deduced from the fragments obtained and for already known NDP-glucose dehydratases. From these studies it became evident that NDP-glucose dehydratases from Actinomycetes are more closely related to each other than to dehydratases from species of other orders. This phylogenetic analysis also revealed a close relationship between dehydratases from strains that produce natural compounds with similar deoxysugar motifs (Fig. 2) [130]. These results clearly show that dehydratase

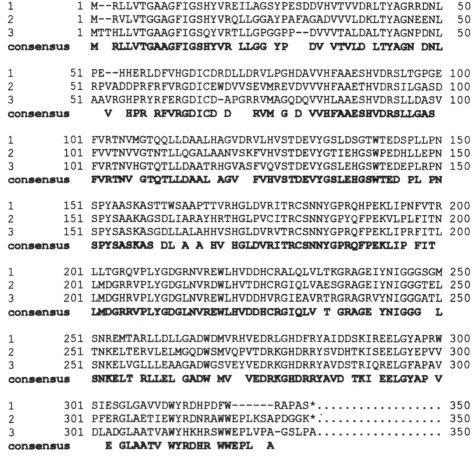

```
1          1 M--RLLVTGAAGFIGSHYVREILAGSYPESDDVHVTVVDRLTYAGRRDNL  50
2          1 M--RVLVTGGAGFIGSHYVRQLLGGAYPAFAGADVVVLDKLTYAGNEENL  50
3          1 MTTHLLVTGAAGFIGSQYVRTLLGPGGPP--DVVVTALDALTYAGNPDNL  50
consensus    M  RLLVTGAAGFIGSHYVR LLGG YP    DV VTVLD LTYAGN DNL

1         51 PE--HHERLDFVHGDICDRDLLDRVLPGHDAVVHFAAESHVDRSLTGPGE 100
2         51 RPVADDPRFRFVRGDICEWDVVSEVMREVDVVVHFAAETHVDRSILGASD 100
3         51 AAVRGHPRYRFERGDICD-APGRRVMAGQDQVVHLAAESHVDRSLLDASV 100
consensus     V  HPR RFVRGDICD D    RVM G D VVHFAAESHVDRSLLGAS

1        101 FVRTNVMGTQQLLDAALHAGVDRVLHVSTDEVYGSLDSGTWTEDSPLLPN 150
2        101 FVVTNVVGTNTLLQGALAANVSKFVHVSTDEVYGTIEHGSWPEDHLLEPN 150
3        101 FVRTNVHGTQTLLDAATRHGVASFVQVSTDEVYGSLEHGSWTEDEPLRPN 150
consensus    FVRTNV GTQTLLDAAL AGV   FVHVSTDEVYGSLEHGSWTED PL PN

1        151 SPYAASKASTTWSAAPTTVRHGLDVRITRCSNNYGPRQHPEKLIPNFVTR 200
2        151 SPYSAAKAGSDLIARAYHRTHGLPVCITRCSNNYGPYQFPEKVLPLFITN 200
3        151 SPYSASKASGDLLALAHHVSHGLDVRVTRCSNNYGPRQFPEKLIPRFITL 200
consensus    SPYSASKAS DL A A HV HGLDVRITRCSNNYGPRQFPEKLIP FIT

1        201 LLTGRQVPLYGDGRNVREWLHVDDHCRALQLVLTKGRAGEIYNIGGGSGM 250
2        201 LMDGRRVPLYGDGLNVRDWLHVTDHCRGIQLVAESGRAGEIYNIGGGTEL 250
3        201 LMDGHRVPLYGDGLNVREWLHVDDHVRGIEAVRTRGRAGRVYNIGGGATL 250
consensus    LMDGRRVPLYGDGLNVREWLHVDDHCRGIQLV T GRAGE YNIGGG  L

1        251 SNREMTARLLDLLGADWDMVRHVEDRLGHDFRYAIDDSKIREELGYAPRW 300
2        251 TNKELTERVLELMGQDWSMVQPVTDRKGHDRRYSVDHTKISEELGYEPVV 300
3        251 SNKELVGLLLEAAGADWGSVEYVEDRKGHDRRYAVDSTRIQRELGFAPAV 300
consensus    SNKELT RLLEL GADW MV   VEDRKGHDRRYAVD TKI EELGYAP V

1        301 SIESGLGAVVDWYRDHPDFW------RAPAS*..................  350
2        301 PFERGLAETIEWYRDNRAWWEPLKSAPDGGK*..................  350
3        301 DLADGLAATVAWYHKHRSWWEPLVPA-GSLPA..................  350
consensus     E GLAATV WYRDHR WWEPL  A
```

Fig. 3. Alignment of the amino acid sequences of *1*: GraE from *S. violaceoruber* (Tü22); *2*: Gdh from *Saccharopolyspora erythraea*; *3*: StrE from *S. griseus*. The amino acids present at least in 2 of 3 are defined as consensus

genes can be used as targets for genetic screening experiments. An overview of the deduced protein sequences of different NDP-glucose dehydratases is depicted in Fig. 3.

3.1.3
Genes and Proteins Involved in the Conversion of NDP-4-keto-6-Deoxyhexose to NDP-3,6-dideoxy-4-keto-Hexose

Two genes are involved in the biosynthesis of CDP-3,6-dideoxy-4-keto-glucose, an intermediate of the biosynthesis of ascarylose in *Yersinia pseudotuberculosis* [53]. As was already shown in Scheme 3 (Sect. 2.1.2.), the conversion of CDP-4-keto-6-deoxyglucose is catalyzed by a dehydrase (E1; gene product of *ascC*). An intermediate is then converted to dCDP-3,6-dideoxy-4-keto-glucose by a NAD(P)H-dependent reductase (E3; gene product of *ascD*). Based on the physical characteristics of AscC and AscD, a molecular mechanism for the C-3 deoxygenation has been proposed. AscC, a PMP-dependent enzyme, produces a PMP-$\Delta^{3,4}$-glucoseen intermediate. The conversion of this compound to the final 3,6-dideoxyhexose is catalyzed by AscD via a radical mechanism [21]. Carefull characterization of AscC and AscD showed that both enzymes contain iron-sulfur cluster motifs, which are important for electron flow during C-3 deoxygenation.

Genes coding for proteins which are homologous to AscC have been found in a few biosynthetic gene clusters coding for antibiotics (see Sect. 3.1.4) [102, 131]. The deduced protein sequences of *rfbH* and *rfbI* from *Salmonella enterica LT2* show particularly strong homology to AscC and AscD, respectively, indicating the same function for these proteins as for AscC and AscD [21].

3.1.4
Genes and Proteins Involved in the Incorporation of Amino Groups

As was just described, several genes coding for proteins showing strong homology to AscC have been detected in antibiotic gene clusters. These genes are *dnrJ, ery C1, tylB, strA, strC, strS* and *prgI* [102, 131b]. Their deduced proteins show modest residue congruence to PLP-dependent aminotransferases [53]. Based on these homologies and on the fact that only a slight mechanistic divergence from the AscC-catalyzed dehydration would lead to a PLP/PMP-dependent transamination, all these gene products were considered to be aminotransferases. It could be demonstrated that StrC is an aminotransferase involved in the biosynthesis of streptidin [103]. All these proteins have a conserved amino acid motif: $Gx_3Dx_7Ax_8EDx_{10}Gx_3Gx_{13}Kx_{4-5}geGGx_{19}G$ (Fig. 4) [102]. The conserved lysine (K), postulated as a pyridoxal-binding residue, is replaced by a histidine (H) in AscC or RfbH. An alignment of AscC, DnrJ and EryC1 is given in Fig. 4.

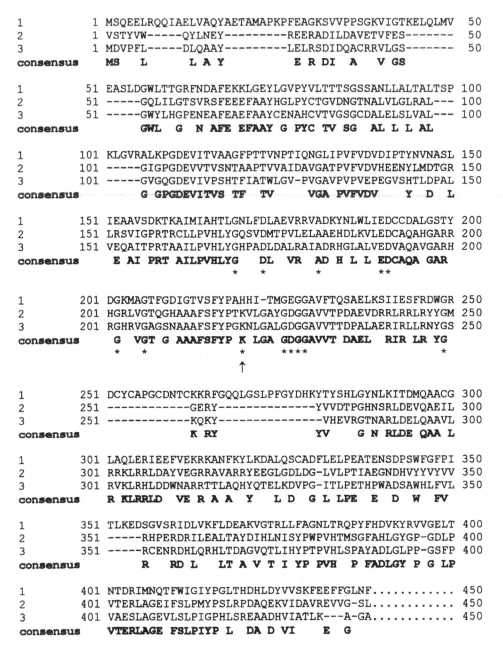

```
1      1 MSQEELRQQIAELVAQYAETAMAPKPFEAGKSVVPPSGKVIGTKELQLMV  50
2      1 VSTYVW-----QYLNEY---------REERADILDAVETVFES-------  50
3      1 MDVPFL-----DLQAAY---------LELRSDIDQACRRVLGS-------  50
consensus  MS  L     LAY        E RDI A   V GS


1     51 EASLDGWLTTGRFNDAFEKKLGEYLGVPYVLTTTSGSSANLLALTALTSP 100
2     51 -----GQLILGTSVRSFEEEFAAYHGLPYCTGVDNGTNALVLGLRAL--- 100
3     51 -----GWYLHGPENEAFEAEFAAYCENAHCVTVGSGCDALELSLVAL--- 100
consensus     GWL  G  N AFE EFAAY G PYC TV SG  AL L L AL


1    101 KLGVRALKPGDEVITVAAGFPTTVNPTIQNGLIPVFVDVDIPTYNVNASL 150
2    101 -----GIGPGDEVVTVSNTAAPTVVAIDAVGATPVFVDVHEENYLMDTGR 150
3    101 -----GVGQGDEVIVPSHTFIATWLGV-PVGAVPVPVEPEGVSHTLDPAL 150
consensus       G GPGDEVITVS TF  TV    VGA PVFVDV    Y D  L
                                  *      *        *       **


1    151 IEAAVSDKTKAIMIAHTLGNLFDLAEVRRVADKYNLWLIEDCCDALGSTY 200
2    151 LRSVIGPRTRCLLPVHLYGQSVDMTPVLELAAEHDLKVLEDCAQAHGARR 200
3    151 VEQAITPRTAAILPVHLYGHPADLDALRAIADRHGLALVEDVAQAVGARH 200
consensus     E AI PRT AILPVHLYG   DL  VR   AD H L L EDCAQA GAR
                          *     *        *       **


1    201 DGKMAGTFGDIGTVSFYPAHHI-TMGEGGAVFTQSAELKSIIESFRDWGR 250
2    201 HGRLVGTQGHAAAFSFYPTKVLGAYGDGGAVVTPDAEVDRRLRRLRYYGM 250
3    201 RGHRVGAGSNAAAFSFYPGKNLGALGDGGAVVTTDPALAERIRLLRNYGS 250
consensus     G  VGT G AAAFSFYP K LGA GDGGAVVT DAEL  RIR LR YG
                  *     *              *     ****                    *
```

↑

```
1    251 DCYCAPGCDNTCKKRFGQQLGSLPFGYDHKYTYSHLGYNLKITDMQAACG 300
2    251 ------------GERY---------------YVVDTPGHNSRLDEVQAEIL 300
3    251 ------------KQKY---------------VHEVRGTNARLDELQAAVL 300
consensus             K RY              YV     G N RLDE QAA L


1    301 LAQLERIEEFVEKRKANFKYLKDALQSCADFLELPEATENSDPSWFGFPI 350
2    301 RRKLRRLDAYVEGRRAVARRYEEGLGDLDG-LVLPTIAEGNDHVYYVYVV 350
3    301 RVKLRHLDDWNARRTTLAQHYQTELKDVPG-ITLPETHPWADSAWHLFVL 350
consensus  R KLRRLD  VE R A  Y   L D  G L LPE   E  D W  FV


1    351 TLKEDSGVSRIDLVKFLDEAKVGTRLLFAGNLTRQPYFHDVKYRVVGELT 400
2    351 -----RHPERDRILEALTAYDIHLNISYPWPVHTMSGFAHLGYGP-GDLP 400
3    351 -----RCENRDHLQRHLTDAGVQTLIHYPTPVHLSPAYADLGLPP-GSFP 400
consensus       R  RD L   LT A V T I YP PVH   P FADLGY P G LP


1    401 NTDRIMNQTFWIGIYPGLTHDHLDYVVSKFEEFFGLNF............ 450
2    401 VTERLAGEIFSLPMYPSLRPDAQEKVIDAVREVVG-SL............ 450
3    401 VAESLAGEVLSLPIGPHLSREAADHVIATLK---A-GA............ 450
consensus  VTERLAGE FSLPIYP L  DA D VI     E  G
```

Fig. 4. Alignment of the amino acid sequences of *1*: AscC from *Yersinia pseudotuberculosis*; *2*: DnJ from *S. peucetius*; *3*: EryC1 from *Saccharopolyspora erythraea*. The amino acids present at least in 2 of 3 are defined as consensus. A postulated amino acid motif is given as *. ↑ indicates a possible pyridoxal-binding site in CnrJ and EryC1

3.1.5
Genes and Proteins Involved in Epimerization

Genes coding for possible epimerases have been detected in several gene clusters and some of them have been expressed in heterologous systems. dTDP-rhamnose was enzymatically synthesized from dTDP-glucose using extracts of *E. coli* harboring plasmids containing different parts of the *rfb* gene cluster of *Salmonella enterica LT2*. In theses studies *rfbC* was identified as the gene coding for an epimerase and *rfbD* as coding for a reductase [132]. A similar reaction has been shown to take place in the biosynthesis of CDP-ascarylose in *Y. pseudotuberculosis* and CDP-abequose and CDP-paratose in different *Salmonella* species [21].

The biosynthesis of CDP-tyvelose from CDP-paratose is catalyzed by a C-2 epimerase [133, 134], the gene product of *rfbE*. Other epimerases (StrM, StrF, StrG, StrP, StrX) are suggested to be involved in the biosynthesis of streptomycin (**24**). The functions of these genes have been deduced from sequence similarities to known epimerases [72]. In addition, the gene product of *strM* was used in incubation reactions together with *RfbD*, affording L-rhamnose [103]. In spite of the studies described above, the exact mechanism of epimerization has yet to be elucidated. With the exception of an epimerase catalyzing the transformation of GDP-4-keto-6-deoxy-D-mannose to GDP-L-fucose [135], it appears that commonly two proteins are required for this biosynthetic step in deoxysugar formation. When either AscE or AscF, both of which are involved in the biosynthesis of ascarylose, was used alone in an incubation reaction, no conversion of CDP-3,6-dideoxy-4-keto-D-glucose to CDP-ascarylose was observed. However, by employing the two enzymes together, CDP-ascarylose was produced [21]

3.1.6
Genes and Proteins Involved in Keto Reduction

Stereospecific reductases are involved in the reduction of keto groups to secondary alcohols the desired endproducts of a deoxysugar biosynthesis. This has been shown for the biosynthesis of L-rhamnose and CDP-ascarylose (see Sect. 3.1.4). The reduction of keto groups at C-4 have been proposed to be catalyzed by DnrV (daunorubicin **29**), EryBIV (erythromycin **14**), and StrL (streptomycin **24**) [21].

The function of the gene product of *rdmF*, located in the rhodomycin biosynthetic gene cluster from *S. purpurascens* [136], is still not known. Interestingly, this protein shows significant homology to a glucose-fructose oxidoreductase from *Zymomonas mobilis* (Fig. 5) [137, 138]. This protein is remarkable in that it is able to catalyze the oxidation of glucose to gluconolactone and the concommitant reduction of the second substrate fructose to sorbitol. During this process the pyridine nucleotide cofactor remains bound to the protein (Fig. 6). The significant homology of RdmF with this enzyme strongly indicates that RdmF is involved in oxidation and/or reduction processes in the biosynthesis of sugar intermediates of the anthracycline antibiotic rhodomycin.

```
1     1 IPMPEH------------------RQQRALRMGVIGTANIAIRRIMPV  50
2     1 ATLPAGASQVPTTPAGRPMPYAIRPMPEDRRFGYAIVGLGKYALNQILPG  50
consensus     P                      R    G    A   I P

1    51 LAAHDHVDLVAVASRDKARAERVGAAFGCG-----GVGDYAALVERDDLD 100
2    51 FAGCQHSRIEALVSGNAEKAKIVAAEYGVDPRKIYDYSNFDKIAKDPKID 100
consensus     A  H    A S     A V A G                      D

1   101 AVYIPLPPGMHHEWALRALRSGKHVLVEKPMSDTYEKTLELMSTASELGL 150
2   101 AVYIILPNSLHAEFAIRAFKAGKHVMCEKPMATSVADCQRMIDAAKAANK 150
consensus     AVYI LP   H E A RA   GKHV  EKPM           A

1   151 VLAENFMFLHHS-QHAAVRAMLDESVGELRLFSGSFAVPPLAPESFR--- 200
2   151 KLMIGYRCHYDPMNRAAVKLIRENQLGKLGMVTTD-NSDVMDQNDPAQQW 200
consensus     L              AAV        G L

1   201 -YQPALGGGALLDVGVYPLRAAQLYLTGELDVLGACLRVDETTGA---DV 250
2   201 RLRRELAGGGSLMDIGIYGLNGTRYLLGEEPIEVRAYTYSDPNDERFVEV 250
consensus         L GGG L DG Y L    YL GE                   V

1   251 AGSVLLSDDRGVTAQLDFGFEHSYRST----YALWGNRGRVSVRRAFTPP 300
2   251 EDRIIW-QMRFRSGALSHG-ASSYSTTTTSRFSVQGDKAVLLMDPA---T 300
consensus         R    L G  SY  T        G         A

1   301 EQLKPVVRIE-----QQDVVTERSLPEDNQVFNAMDAFVRAALTNTGPLT 350
2   301 GYYQNLISVQTPGHANQSMMPQFIMPANNQFSAQLDHLAEAVINNKPVRS 350
consensus                  Q        P  NQ     D    A   N

1   351 DTSAIQRQALLLDRV---RRAARLIGDPKSRPHDVLAQDTGLS*  .... 400
2   351 PGEEGMQDVRLIQAIYEAARTGRAVNTDWGYVRQGGY         .... 400
consensus             L       R R         V  D L
```

Fig. 5. Alignment of the amino acid sequences of *1*: RdmF from *S. purpurascens* and *2*: GFOR from *Zymomonas mobilis*

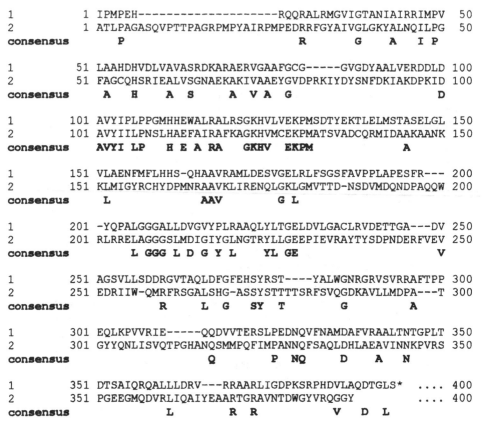

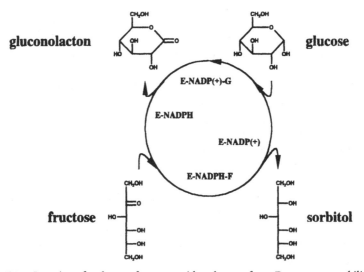

Fig. 6. Reaction of a glucose-fructose oxidoreductase from *Zymomonas mobilis*

3.1.7
Genes Coding for Glycosyltransferases

Only a few genes coding for glycosyltransferases have been identified in antibiotic gene clusters so far. For example, in *Streptomyces lividans* the function of the Mgt protein product is transfer of mycosaminosyl or desosaminyl residues to the aglycone of macrolide antibiotics. The deduced protein sequences of several other genes identified thus far show low but significant homology to Mgt. Genes coding for glycosyltransferases have been detected in the biosynthetic gene clusters of the antibiotics daunomycin (**29**) [139, 140], amphotericin [103], granaticin (**10**) [122], oleandomycin [141], urdamycin A (**51**) and landomycin A (**12**) [122b]. The degree of identity between the corresponding glycosyltransferases was also found to be significant.

Several glycosyltransferase genes in the gene clusters of *S. enterica* O-antigen (serogroup B, C2, E1) biosynthesis have been reported. In theses studies mannosyl- and rhamnosyl transferase genes were identified and characterized [142]. Mannosyl transferases and rhamnosyl transferases form two separate families, but there was no relation between them nor were any conserved motifs.

3.2
Gene Clusters Involved in the Biosynthesis of Deoxysugar Moieties of Antibiotics and O-Antigen Specific Polysaccharides

In recent years, genes coding for proteins involved in sugar biosynthesis have been cloned from several antibiotic-producing microorganisms (Table 1).

Table 1. List of antibiotics and strains from which genes involved in deoxysugar biosynthesis have been cloned

Antibiotic	Strain	Reference
Amphotericin	*S. nodosus*	103
Avilamycyn	*S. viridochromogenes Tü57*	130
Chlorothricin	*S. antibioticus Tü57*	122 c
Daunomycin	*S. C5*	139
	S. peucetius	140, 142
	S. griseus JA3933	147
Erythromycin	*Saccharopolspora erythraea*	24, 146
Granaticin	*S. violaceoruber Tü22*	122 a, d
Hydroxystreptomycin	*S. glaucescens*	72
Landomycin	*S. cyanogenus S136*	130
Lincomycin	*S. lincolnenis*	145
Rhodomycin	*S. purpurascens*	136
Streptomycin	*S. griseus N2-3-11*	72
Tylosin	*S. fradiae*	143
Urdamycin	*S. fradiae Tü2717*	130

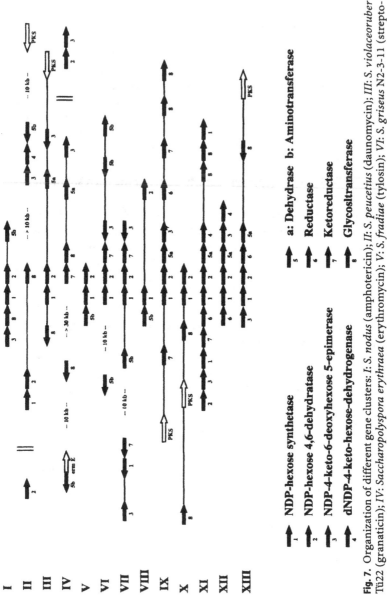

Fig. 7. Organization of different gene clusters: *I*: *S. nodus* (amphotericin); *II*: *S. peucetius* (daunomycin); *III*: *S. violaceoruber* Tü22 (granaticin); *IV*: *Saccharopolyspora erythraea* (erythromycin); *V*: *S. fradiae* (tylosin); *VI*: *S. griseus* N2-3-11 (streptomycin); *VII*: *S. glaucescens* (hydroxystreptomycin); *VIII*: *S. lincolnensis* (lincomycin); *IX*: *S. cyanogenus* S-136 (landomycin A); *X*: *S. fradiae* Tü2717 (urdamycin A); *XI*: *Salmonella enterica* LT2 type B (O antigen); *XII*: *Yersinia pseudotuberculosis*, O antigen; *XIII*: *S. antibioticus* Tü99. *Parallel lines indicate that genes are not clustered*

NDP-hexose synthetase

NDP-hexose 4,6-dehydratase

NDP-4-keto-6-deoxyhexose 5-epimerase

dNDP-4-keto-hexose-dehydrogenase

a: Dehydrase b: Aminotransferase

Reductase

Ketoreductase

Glycosyltransferase

As with other antibiotic biosynthesis genes, these genes show a tendency to cluster. The arrangement of ORFs coding for proteins involved in deoxysugar biosynthesis of several antibiotics or lipopolysaccharides is presented in Fig. 7.

Genes coding for NDP-hexose synthetases and NDP-hexose 4,6-dehydratases have been found in the clusters of the antibiotics granaticin (**10**) [122], streptomycin (**24**) [72], hydroxy-streptomycin [72], amphotericin [103], tylosin (**46**) [143], daunomycin (**29**) [144], lincomycin [145], landomycin A (**12**) and urdamycin A (**51**) [122b, 130]. In most cases, the stop codon of NDP-hexose synthetase genes overlaps the start codon of the NDP-hexose dehydratase genes suggesting that these genes form a transcription unit.

However, the dehydratase gene is not always located in the biosynthetic gene cluster. The dehydratase gene *gdh* has been isolated from the erythromycin-producing strain *Saccharopolyspora erythraea*. Southern analysis indicated that *S. erythraea* contains only one copy of this gene. Surprisingly, this gene was not located within or near the known boundaries of the erythromycin cluster, while other genes governing the deoxysugar portion of the erythromycin biosynthetic pathway were shown to lie up to 15 kb upstream or downstream from the erythromycin resistance gene *ermE* [24, 146].

The dehydratase gene *dnrM* from *S. peucetius* is located in the daunorubicin biosynthetic cluster. Although the gene was believed to be required for the synthesis of daunosamine, a frameshift in the DNA sequence was detected which causes premature termination of translation. Inactivation of *dnrM* did not prevent daunorubicin production. The data indicate that the product of a second dehydratase gene detected outside of the biosynthetic gene cluster is involved in the biosynthesis of daunomycin (**29**) [144].

In all other cases investigated so far, the entire set of biosynthetic genes is located in one region of the chromosome. However, in contrast to polyketide biosynthetic genes, biosynthetic genes coding for sugars are not always tightly grouped in these clusters. In some cases biosynthetic genes for the same deoxysugar are distributed over a distance of more than 10 kb from each other.

The 3,6-dideoxyhexose biosynthetic genes from *Y. pseudotuberculosis* (*ascD, ascA, ascB, ascC, ascE, ascF*) and *S. enterica* LT2 (serogroup B, C2) (*rfbI, rfbF, rfbG, rfbH, rfbS, rfbE*) provide another example of genes clustered on the chromosome. Based on the G/C content within these clusters, the presence of gene sets (for example, in *Yersinia pseudotuberculosis*: set 1: *ascD*; set 2: *ascABC*; set 3: *ascEF*) with different origins or histories has been suggested. Sets of genes with similar function in both strains show a similar G/C content, indicating either that the clusters were incorporated into both strains at the same time in evolution or that recombination events between the two strains must have occurred [21].

4
Contribution of Mono- or Oligosaccharide Moieties to Biological Activity

Although carbohydrate-containing antibiotics and other bioactive natural products have been known for decades [148], except for the carbohydrate-based glycosidase inhibitors (see Sect. 2.1.5) little research has been devoted to the biological or pharmacological role of these sugar moieties. Clearly, the aglycon itself is not active in most instances, as was demonstrated for many antibiotics and antitumor compounds, erythromycin (**14**), daunomycin (**29**) and amphotericin B being prominent examples. Traditionally, the glycan chains of these glycoconjugates have been viewed as molecular elements that control the pharmacokinetics of a drug, such as absorption, distribution, metabolism and excretion. This notion, however, is beginning to change. The rigid character of the pyran rings along with the flexibility associated with the glycosidic linkages give them the ability of preorganization. A delicate balance between hydrophilic and hydrophobic domains is an additional feature of deoxygenated carbohydrates. Recent results clearly show that deoxygenated oligosaccharides in natural products from microbial sources can actively contribute as recognition elements to the mechanism of action of the respective drugs.

By far one of the best targets for testing interactions of a drug with its receptor is DNA. Firstly, many carbohydrate-bearing antitumor antibiotics are known to act at the DNA level and therefore a contribution of the deoxysugar moiety to cytostatic activity can be traced either to faciliated cell membrane permeability or to adhesion, i. e., binding to the DNA itself. Secondly, any self-complementary DNA fragment (palindrome) can be synthesized in amounts sufficient for experimental studies. Finally, due to DNA's helical and relatively stable conformation in solution, DNA-drug complexes can conveniently be studied by high resolution NMR-spectroscopy [149]. In contrast, no data on the role of carbohydrates in antibiotics that operate at the ribosomal level, such as erythromycin (14), are available so far.

Structural evidence for a direct interaction between duplex DNA and a carbohydrate moiety of a gylcoconjugate came from crystallographic studies [150]. For example, daunomycin (29) was co-crystalized with the hexanucleotides d(CGATCG) and d(CGTAGC) and, consistent with similar cases [151c], the crystallographic data unequivocally showed that the aminosugar daunosamine (see Scheme 28, 63) is located in the minor groove of the DNA duplex [151]. Here, the protonated amino group strongly contributes to the noncovalent binding of daunomycin to the DNA fragment. Detailed analysis revealed hydrogen

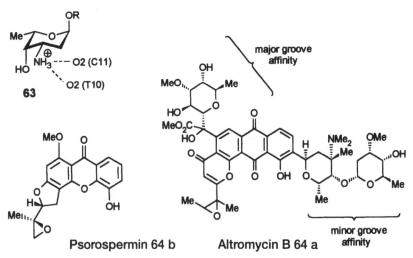

Scheme 28. The aminosugar daunosamine (63) is located in the minor groove of the DNA duplex. The disaccharide form of altromycin B (64a)is located in the minor groove of the DNA palindrome. A parallel orientation of the xanthone ring structure to adjacent base pairs is favored by psorospermin (64b)

bonding between the NH_3^+ group and the O-2 of cytidine and a thymidine, respectively. Interestingly, electrostatic interactions between the negatively charged phosphate backbone and the positive charge on the trideoxy aminosugar were not detected. In this daunosamine/DNA complex, the aminosugar, unlike its rigid aglycon, is subject to a pronounced conformational distortion.

Altromycin B (**64a**), a member of the pluramycin family of antibiotics, is an antitumor agent which induces DNA cleavage. An important feature of **64a** is the two deoxysugar domains, both of which are attached via C-glycosidic linkages to the chromophore. Hurley and coworkers conducted NMR studies on the altromycin/DNA-complex [152a, b]. These studies in solution indicated that the disaccharide portion is located in the minor groove of the DNA palindrome [d(GAAGTACTTC)$_2$] while the monosaccharide moiety is found in the major groove. The authors proposed that the orientation and binding of the glycan units are crucial for the observed DNA sequence specificity of altromycin B. Further support was gained from NMR studies of the psorospermin/DNA complex [152c]. Psorospermin **64b** is a cytotoxic, DNA strand-breaking dihydrofuranoxanthone which, except for the missing deoxysugars, shows structural similarities to altromycin B **64a**. However, instead of perpendicularly intercalating with the axis of the tetracyclic chromophore between two base pairs, like **64a** does, a parallel orientation of the xanthone ring structure to adjacent base pairs is favored in **64b**, again demonstrating the role of carbohydrates in DNA-binding glycoconjugates. Related studies on the anthracycline antibiotics nogalamycin [153] and menogaril [154] led to similar observations and further supported the role of charged aminodeoxysugars in noncovalent binding in duplex DNA.

By far the best studied glycoconjugate in this respect is calicheamicin γ_1^I (**65**), a member of the enediyne antibiotics, which possesses high antitumor activity [155]. It consists of an aglycon core with a 1,5-diyn-3-ene unit, an allylic methyl trisulfide and an α,β-unsaturated ketone, all of which are important for the ability of **65** to cleave both strands of DNA with a high degree of sequence specificity and hence to cause cell death. Furthermore, calicheamicin γ_1^I contains a deoxyoligosaccharide chain **66** (depicted as the methyl glycoside) with an array of unusual and novel structural features. In NMR studies on the DNA/drug complex of **65**, Kahne et al. [156] and others [157] found that the oligosaccharide unit is well embedded in the minor groove of duplex DNA, thereby positioning the enediyne moiety for cleavage.

Nicolaou [158] and Danishefsky [159] and their coworkers went one step further by synthesizing the methyl glycoside of the calicheamicin oligosaccharide (see Sect. 5.1.2.1), thereby paving the way for studies on DNA/glycan complexes of glycoconjugates without interference by the aglycon. Various "contacts" between DNA and glycan chain atoms were established by NMR-spectroscopy, most notably, as had been postulated from molecular dynamic calculations [160], between the polarizable aromatic iodine and the exocyclic C-2 NH_2 groups of two guanine residues [161]. Changes of the chemical shifts ($\Delta\delta$) of DNA protons induced by calicheamicin Θ [162] were compared with the corresponding changes generated by the methyl glycoside **66** and were found to be close to identity. Furthermore, the aglycon calicheamicinone shows little affinity to DNA and largely lacks binding specificity to the four sequences TCCT, TTTT, TCTC, and ACCT, which is a unique feature of **65** [163]. A fascinating property of the methyl glycoside **66** is its ability to inhibit DNA binding of a transcription factor whose recognition sequence includes the calicheamicin binding site. Furthermore, **66** inhibits in vivo expression of a reporter gene controlled by that transcription factor, suggesting a strategy for the development of a class of novel biological probes and therapeutic agents [164]. From all these results it was concluded that the primary role of the sugar domain of **65** is recognition and binding ($K_D \sim 10^{-6}$) in the minor groove of duplex DNA, with a high degree of sequence specificity.

As an extension of these studies, a tethered head-to-head dimer **67** was designed [165] which exhibits a much higher affinity to duplex DNA recognition sites ($K_D = 10^{-9}$) and shows a higher degree of sequence specificity [166]. Furthermore, **67** inhibits binding of transcription factors to DNA with a significantly higher activity than **66** [167]. NMR studies with the B form duplex d(CGT*AGGA*TA*TCCT*ACG)$_2$ revealed that the head-to-head dimer-DNA complex showed the same kind of contacts detected in the corresponding complex with **65** [168].

All these data suggest that carbohydrates are potentially able to generate DNA target specificity and may serve to control the function of nucleic acids. It is reasonable to assume that the affinity of a DNA-binding antibiotic for DNA may be determined by the structure and length of the sugar chain attached to the aglycon [165, 169].

5
Synthetic Aspects of the Chemical and Enzymatic Construction of Deoxygenated Oligosaccharides

The increased understanding of the role of carbohydrates in nature and the subsequent impact on the biological and pharmaceutical sciences has led to a renewed interest in synthetic carbohydrate chemistry. Until recently, construction of oligosaccharides composed of deoxygenated hexoses has relied exclusively on chemical methods [10, 170]; lately, however, many elegant improvements have been made. In particular, the development of strategies for a one-pot reaction multistep construction of tri- and larger oligosaccharides, by exploiting the armed/ disarmed principle [171], and the solid phase synthesis of oligosaccharides [172] are currently topics of great interest. With the improved understanding of the biosynthesis of secondary metabolites, including the deoxyhexoses, and better access to the corresponding enzymes using genetic engineering techniques, enzymatic oligosaccharide synthesis, including the preparation of deoxygenated glycosides, has emerged as a practical alternative to chemical synthesis. The use of glycosyltransferases avoids protection/deprotection sequences and usually guarantees a high stereo- and regioselectivity. However, in most cases nucleotide sugars are required as glycosyl donors for glycosyltransferases and the availability of the donor substrate and the corresponding enzyme is still the limiting factor in enzyme-mediated oligosaccharide synthesis. Nevertheless, biocatalysts involved in carbohydrate metabolism, such as phosphoglucomutase

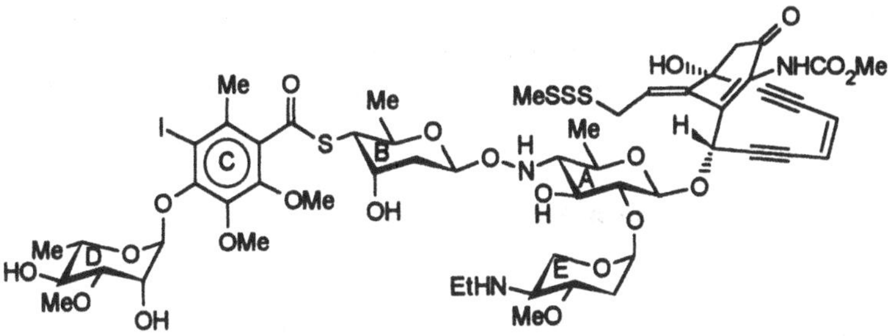

Calicheamicin γ¹₁ 65

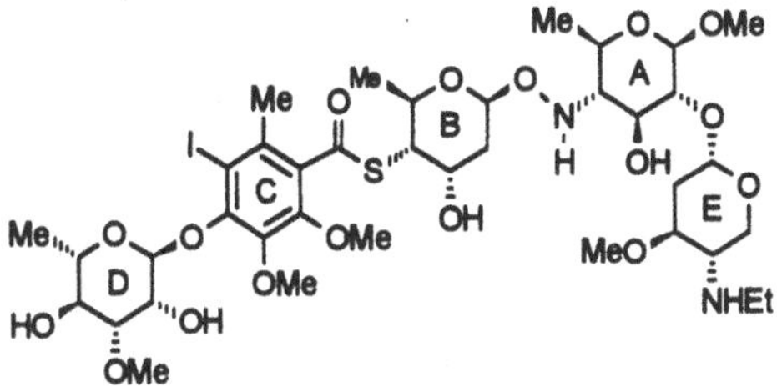

66

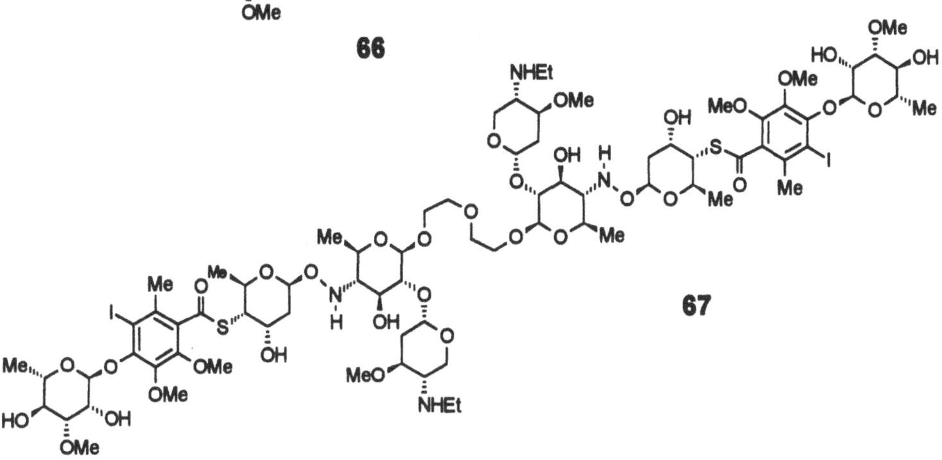

67

Scheme 29. Calicheamicin γ_1^1 (**65**) and its methyl deoxyoligosaccharide chain (**66**); the tethered head-to-head dimer (**67**)

and uridine diphosphoglucose pyrophosphorylase [32], dTDP-D-glucose-4,6-dehydratase [173], and sucrose synthase [174] have recently been employed for constructing deoxygenated sugar nucleotide diphosphates in a preparative scale. The use of these compounds in the synthesis of oligosaccharides in the presence of suitable glycosyltransferases is only beginning to be explored [175].

It is the intention of this section to present recent advances in the the art of modern oligosaccharide synthesis, with particular emphasis on methods for the stereoselective assembly of 2-deoxyglycosides. The strengths of the various chemical methodologies will be discussed in a selected number of examples.

Additionally, it will be shown that the use of enzymes holds great promise both for the enzymatic or chemoenzymatic construction of deoxygenated donor substrates of glycosyltransferases and for enzymatic oligosaccharide synthesis.

5.1
Chemical Glycosidation with Deoxysugars

5.1.1
Techniques for the Synthesis of 2-Deoxy-α- and β-Glycosides

Glycosidation is one of the key reactions of life. As in nature, chemists usually perform glycosidations by abstraction of appropriate leaving groups at the anomeric carbon using specific promoters or catalysts in the presence of a glycosyl acceptor [170]. Chemical synthesis of 2-deoxyglycosides has been a particular challenge in oligosaccharide synthesis. The main problem associated with the construction of 2-deoxyglycosides is the absence of a neighboring group at C-2 that is able to contribute stereodirecting achimeric assistance in the glycosidation step. In addition, due to the lower stability of the glycosidic bond, 2-deoxyglycosides are more difficult to handle; this is particularly true for 2-deoxyglycosyl halides. Therefore, conventional Koenigs-Knorr glycosidation conditions have rarely been employed [176]. Alternatively the use of, more stable leaving groups, e.g., thioethers in thioglycosides **68** and fluorine, present in glycosyl fluorides **69**, are examples of important recent achievements (Scheme 30). Both glycosyl donors were originally utilized for oligosaccharide synthesis with "conventional" sugars such as glucose and galactose but have lately found use in 2-deoxysugar chemistry [177]. An important advantage of

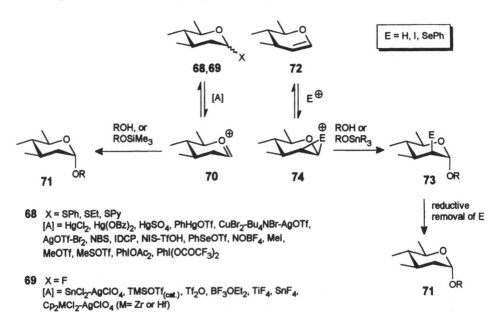

68 X = SPh, SEt, SPy
[A] = HgCl$_2$, Hg(OBz)$_2$, HgSO$_4$, PhHgOTf, CuBr$_2$-Bu$_4$NBr-AgOTf, AgOTf-Br$_2$, NBS, IDCP, NIS-TfOH, PhSeOTf, NOBF$_4$, MeI, MeOTf, MeSOTf, PhIOAc$_2$, PhI(OCOCF$_3$)$_2$

69 X = F
[A] = SnCl$_2$-AgClO$_4$, TMSOTf$_{(cat.)}$, Tf$_2$O, BF$_3$OEt$_2$, TiF$_4$, SnF$_4$, Cp$_2$MCl$_2$-AgClO$_4$ (M= Zr or Hf)

Scheme 30. Strategies for the chemical synthesis of 2-deoxy-α-glycosides thioglycosides is that they provide starting material for glycosyl fluorides. Activation of these functional groups and formation of intermediate oxonium ion **70** can be achieved by very specific reagents. Thioglycosides are typically activated by metal salts based on palladium [178], copper [179], silver [180], mercury [181] or combinations of these promoters [182], or by positively polarized selenium [183], bromonium [184] or iodonium ions [185], alkyl compounds [186] as well as hypervalent iodine reagents [187]. Typical fluorophilic reagents are Lewis acids based on boron [188], tin [189], silicon [190], titanium [191], zirconium [191], [192], hafnium [192–194] and gallium [196], in some cases in combination with silver salts [192–195]. To a lesser extent anomeric acetates (X = OAc) [197], p-nitrobenzoates (X = pNO$_2$BzO) [198], imidazolethiocarbonates (X = OC(S)imid) [199], S-xanthates (SC(S)OEt) [200], silyl ethers (X = OSiR$_3$) [201] and phosphites (X = OP(OCH$_2$CCl$_3$)$_2$, X = OP(OR)$_2$) [202] have been employed for the construction of 2-deoxyglycosides. When no participating group at C-2 is present, with the exception of the triethyl phosphites [202b] which favor formation of β-2-deoxyglycosides, all these methods give anomeric mixtures, with α-glycosides **71** predominating. Glycals **72** can be regarded as classic glycosyl donors for the construction of complex 2-deoxyglycosides [205a] and their use in this field is a continuing success story [170g, h]. Just recently, glycals were employed by Danishefsky and coworkers for the synthesis of conventional oligosaccharides in solution [170g, h, 203a] and on solid phase [170h, 203b]. In the synthesis of 2-deoxyglycosides, glycals have typically been activated by various electrophiles (E = H [204], I [205], Br [206] and SePh [208]). In the presence of an alcohol or stannyl ether [208] they afford 1,2-*trans* configurated glycosides **73** (E ≠ H). Here, the electrophile regio- and stereoselectively adds to the electron-rich enol ether double bond, leading to the intermediate oxonium ion **74**. Thus, this strategy usually furnishes α-glycosides **73** with high selectivity which may be further transformed into 2-deoxy-α-glycosides **71** by reductive removal of E.

A novel and unique glycosidation strategy of glycals uses dienes with terminal heteroatoms in cycloadditions with glycals [209]. For example, diacylthione **75** regio- and stereoselectively adds to tri-O-benzyl-D-glucal **76** to give cycloaddition product **77** (Scheme 31) [210]. After reductive removal of the sulfur atom by Raney Ni, the 2-deoxy-α-glycoside **78** becomes accessible. Although yields are only moderate for both steps, this approach will no doubt eventually prove to be useful.

Regarding the synthesis of 2-deoxysugars, selective β-glycosidation is still a major challenge. Typically, it is achieved by a neighboring group which is either a structural element of the glycosyl donor or is temporarily placed in the molecule. In fact, electron-donating axial substituents at C-3, e.g., in benzoic ester **79** or urethanes, can stabilize a charged anomeric center, as depicted in the 1,3-acyloxonium ion **80,** thus blocking the α-face in the glycosidation step (Scheme 32) [181d, 211]. 2-Deoxy-β-glycosides **81** are predominantly formed, but this approach highly depends on the configuration at C-3 [212].

A number of alternative and more general approaches have been developed by Thiem and coworkers [213] as well as by the groups of Beau [214], Ogawa [215], Frank [216], Nicolaou [189] and van Boom [217]. All of their methods rely

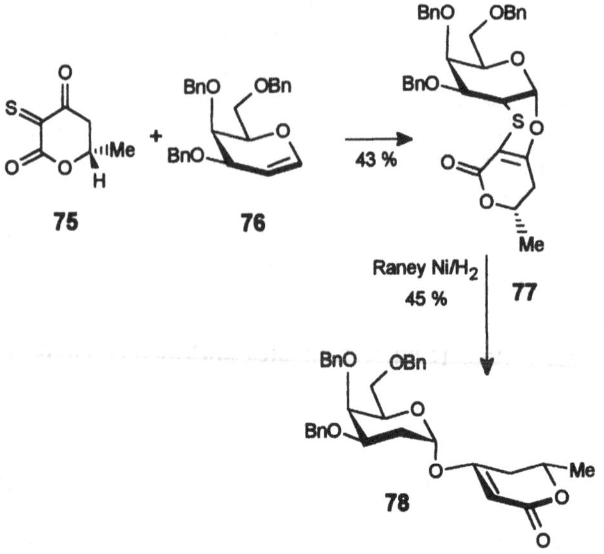

Scheme 31. Preparation of 2-deoxyglycosides using a cycloaddition strategy

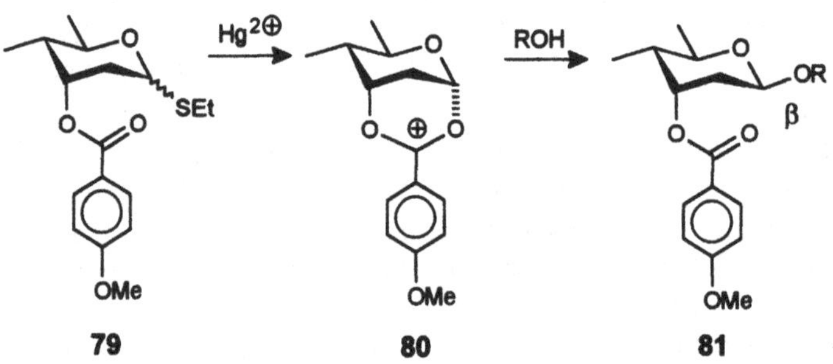

Scheme 32. 2-Deoxy-β-glycosides by 1,3 achimeric assistance

on an equatorially installed C-2 heteroatom substituent that guides the stereo-chemical course of the glycosidation by achimeric participation and which is reductively removed to give β-linked 2-deoxysugars. In the method introduced by Thiem et al., stereoselective introduction of two bromine atoms at C-1 and C-2 is the key step. 2-Bromo-2-deoxy-glycosyl bromides **83** are formed from acetals such as **82** (Scheme 33), and their glycosidation using AgOTf as a promoter predominantly yields β-glycosides **84**. Deblocking at O-3 and reductive debromination gives 2-deoxy-β-glycosides **85**, as was successfully demonstrated in the synthesis of the aureolic acid oligosaccharides (see also mithramycin (**34**); Sect. 2.2.) [213f].

Beau and coworkers prepared 1,2-*trans*-seleno acetates by addition of PhSeCl to electron-rich glycals **71** in the presence of AgOAc (Scheme 34; route A) [214]. Subsequently, these acetates **86** selectively gave β-glycosides **87a** in the presence of TMSOTf as a promoter. Ogawa [215] and Frank [216] and their coworkers

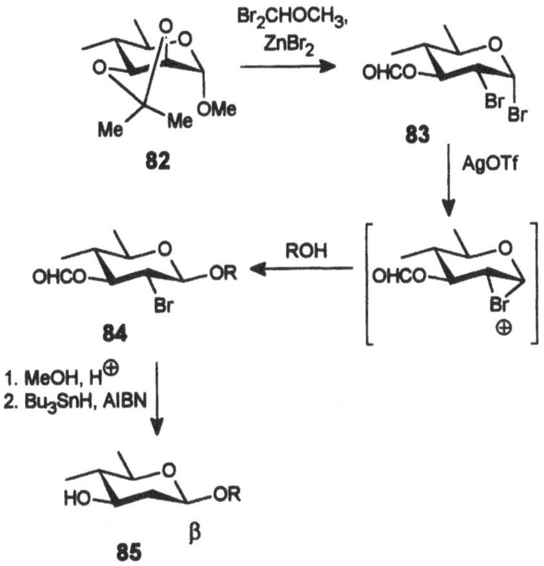

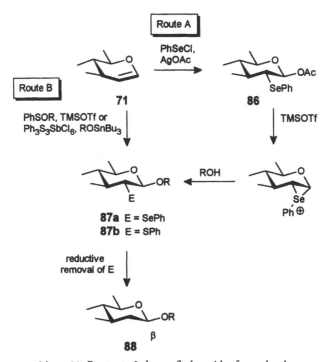

Scheme 33. Br as the stereodirecting group in the preparation of 2-deoxy-β-glycosides

Scheme 34. Routes to 2-deoxy-β-glycosides from glycals

achieved a selective 1,2-*trans* addition to glycals in one step by using phenyl-sulfenate esters in the presence of TMSOTf and by employing phenylbis-(phenylthio)sulfonium salts and an alcohol or stannyl ether preferentially affording 2-thio substituted β-glycosides **87b** (Scheme 34; route B). Again, 2-deoxy-β-glycosides **88** were obtained after reductive removal of the seleno-phenyl and thiophenyl substituent at C-2.

Nicolaou et al. showed that phenylthioglycosides **89a** undergo facile and highly selective 1,2-migration in the presence of diethylamino sulfur trifluoride (DAST) affording 2-deoxy-2-phenylthioglycosyl fluorides **90** (Scheme 35). SnCl$_2$-pro-moted glycosidation in dichloromethane preferentially gives α-glycosides, while under the same conditions the use of diethyl ether as solvent predominantly affords the corresponding β-glycosides **87b** [189]. Improvements in the selec-tivity of this process have been achieved by applying complex mixtures of pro-moters (SnCl$_2$-Cp$_2$ZrCl$_2$-AgOTf) [218]. Similarly, in the presence of NIS/TfOH, phenyl-2-O-(phenoxythiocarbonyl)-1-thioglycosides **89b** undergo 1,2-migra-tion of the phenylthio group to afford 2-phenylthio-2-deoxy-β-glycosides **87b** in one step, as was reported by van Boom and coworkers (Scheme 35) [217].

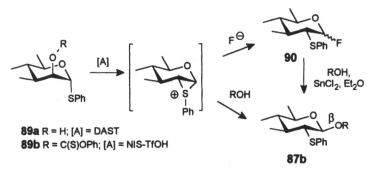

Scheme 35. Arylthio-migration of thioglycosides and β-glycosidation

Recently, Tatsuta and Toshima elegantly designed conformationally rigid 2,6-anhydro-2-thio sugars **91a, b** for the stereoselective preparation of 2,6-dideoxy-O-glycosides[219]. These glycosyl donors can be employed in glycosidation protocols without considering the anomeric effect. Thus, ano-meric acetates **91a** preferentially afford 2-deoxy-β-glycosides **92** while the corresponding glycosyl fluorides and phenylthioglycosides **91b** (activation see above) predominantly give 2-deoxy-α-glycosides **93**. After desulfurization using Raney Ni or tributyltin hydride either 2-deoxy-β-**94** or 2-deoxy-α-glycosides **95** are obtained (Scheme 36).

5.1.2
Recent Highlights in the Construction of Deoxygenated Oligosaccharides

In the following, two examples highlight recent developments in this still rapidly growing field.

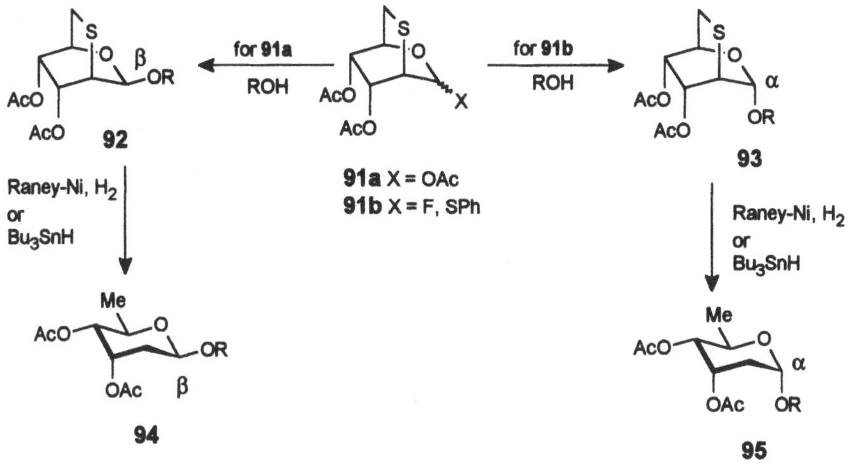

Scheme 36. Glycosidation of 2,6 anhydro-2-thio-sugars

5.1.2.1
Synthesis of the Oligosaccharide Portion of Calicheamicin γ_1^I

An exciting synthetic target in the field of glycoconjugates is calicheamicin γ_1^I **65** (see Scheme 29) [155], whose DNA-binding properties were discussed in Sect. 4. As was pointed out, calicheamicin γ_1^I contains an oligosaccharide chain with an array of unusual and novel structural features such as a β-linked hydroxylamine glycosidic bond, a 2,4-dideoxy-4-amino-L-*threo* pentose (ring E), a sulfur atom at C-4 of ring B and an iodinated, hexasubstituted benzoate (ring C).

Both assembly of the methyl glycoside **66** (see Scheme 29) and coupling of the appropriately protected trichloroacetimidate derivative **96** with the calicheami-cinone γ_1^I precursor **97** [222] (Scheme 37) using Schmidt's protocol [170c] were

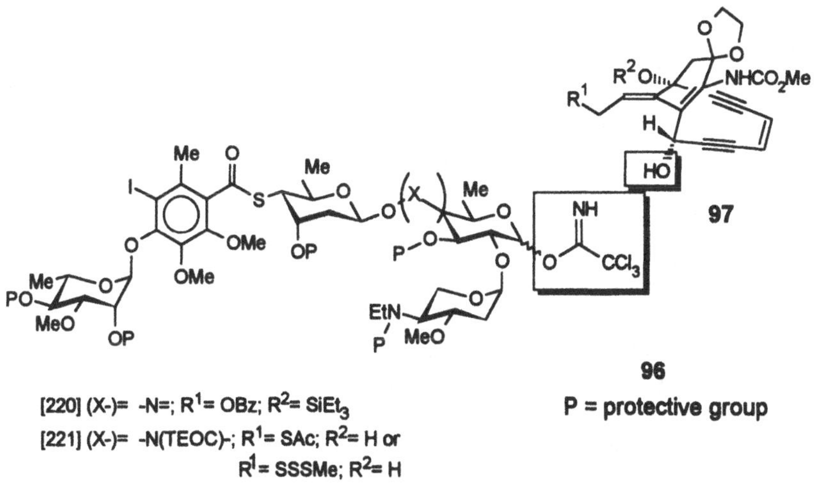

[220] (X-)= -N=; R^1= OBz; R^2= SiEt$_3$
[221] (X-)= -N(TEOC)-; R^1= SAc; R^2= H or
          R^1= SSSMe; R^2= H

P = protective group

Scheme 37. Synthesis of calicheamicin γ_1^I (**65**): The appropriately protected trichloracetimidate derivative (**96**), and the calicheamicinone γ_1^I precursor (**97**) are shown

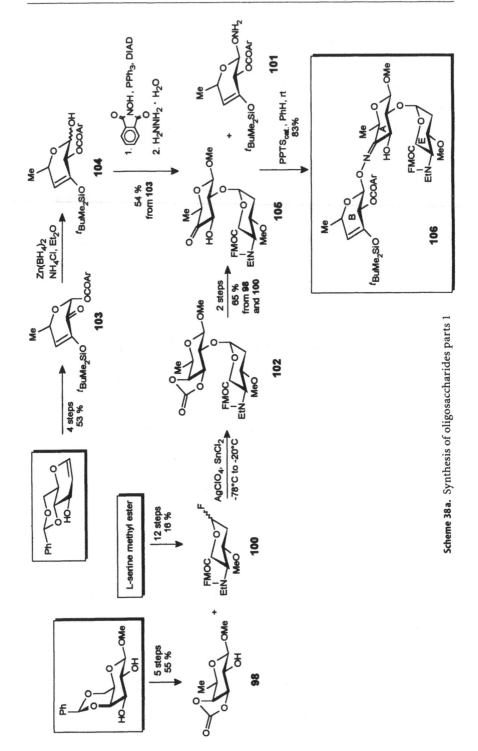

Scheme 38a. Synthesis of oligosaccharides parts 1

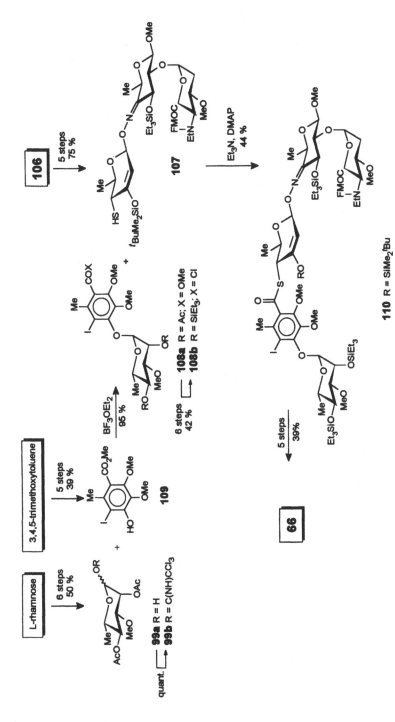

Scheme 38b. Synthesis of oligosaccharides parts 2

achieved by the groups of Nicolaou [220] and Danishefsky [159, 221]. Synthesis of the oligosaccharide domain 66 required application of the entire repertoire of contemporary carbohydrate chemistry, including modern glycosidation methods. Nicolaou's group assembled the oligosaccharide portion of calicheamicin γ_1^I from the bicyclic CD subunit and the BAE ring system [220], but, to set the stage for the key glycosidations, the monomers had to be generated in multistep syntheses. While monosaccharide precursors for rings A 98 (Scheme 38a) and D 99a (Scheme 38b) were obtained by conventional carbohydrate chemistry, construction of the building blocks for rings E 100 and B 101 (Scheme 38a) required development of new synthetic sequences. For example, glycosylfluoride 100 was obtained from L-serine methyl ester, as there was uncertainty about the absolute configuration of the ethylamino substituent when Nicolaou and coworkers started the project and both enantiomers of serine are readily available. B ring intermediate 101 is a 2,6-dideoxy-β-glycoside with a N-O linkage to ring A. However, methodologies using an appropiately placed substituent at C-2 to obtain 2-deoxy-β-glycosides (see Sect. 5.1.1) could not be employed here, due to the potentially problematic reductive removal of this heteroatom substituent in the presence of the easily reducible N-O function appeared. Therefore, the authors prepared enone 103 which after stereoselective reduction and intramolecular acyl migration gave unsaturated pyranose 104 (α:β ca. 7:1). Rapid transfer to the next reaction avoided an epimerization of the anomeric center in 104, and a β-selective introduction of N-hydroxyphthalimide under Mitsunobu conditions, which proceeds through an S_N2-type process with inversion of configuration at the anomeric center, was feasible. The synthesis of 101 (α:β ca. 6:1) is completed by hydrazinolysis of the phthalimide function. 2-Deoxyglycosyl donor 100 was coupled with 6-deoxy-glycoside 98 in the presence of the fluorophilic Lewis acid $AgClO_4$/$SnCl_2$ to give disaccharide 102. Deprotection followed by oxidation afforded 4-ulose 105, which was coupled with the hydroxylaminoglycoside 101 to give oxime 106 (Scheme 38a). A five-step modification of trisaccharide 106 led to thiol 107 which was set for esterification with acyl chloride 108b (Scheme 38b). The latter had been prepared from activated rhamnose 99b and phenol 109 using Schmidt's trichloroacetimidate methodology to afford methyl ester 108a, followed by a sequence of functional group manipulations. Under forcing conditions (Et$_3$N, DMAP$_{cat.}$, CH$_2$Cl$_2$, 0 °C, 1 h) coupling of trisaccharide 107 with benzoyl chloride 108b established pentasaccharide 110. Sequential deprotection, stereoselective reduction of both the B ring and the oxime functionality yielded methyl glycoside 66.

5.1.2.2
Synthesis of the Oligosaccharide Portion of Ciclamycin

Lately, considerable progress has been made in constructing tri- and larger oligosaccharides in one pot [223] by exploiting the armed/disarmed concept. First proposed by Fraser-Reid and coworkers [171], this concept takes into account that the electron-donating/withdrawing power of protective groups in carbohydrates has a profound impact on the reactivitiy of the anomeric center. Electron-poor substituents cause a decrease of reactivity, while electron-rich

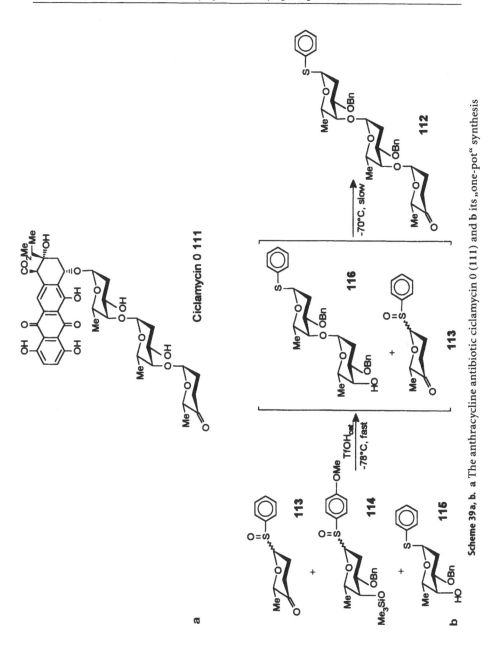

Scheme 39a, b. a The anthracycline antibiotic ciclamycin 0 (111) and b its „one-pot" synthesis

groups enhance it. Thus, tuning of the relative reactivity of glycosyl donors by resident functional groups is possible, and multi-component oligosaccharide synthesis becomes feasible.

A good application of this strategy was reported by Ragharan and Kahne [224] in the synthesis of the trisaccharide of the anthracycline antibiotic, ciclamycin 0 111 (Scheme 39a) [225]. The structure of this highly deoxygenated carbohydrate domain was established through chemical synthesis by Danishefsky and coworkers [226]. While they utilized glycals in a stepwise glycosidation approach, Kahne and coworkers elegantly prepared the trisaccharide 112 by sequential coupling of three monosaccharides in a single reaction. Their one-pot strategy utilized S-glycosides 113–115 which show dramatically different reactivities in glycosidation reactions when activated by triflic acid (Scheme 39b). Here, fine-tuning is achieved by the oxidation level of the sulfur and by the electronic properties imported by the aromatic substituent on sulfur, as well as by manipulation of the glycosyl donor properties (alcohol vs silylether). Thus, the p-methoxyphenylsulfoxide 114 shows the highest degree of reactivity follo-wed by the sulfoxide 113 and the phenyl sulfide 115. Under these conditions the trisaccharide 112 was isolated as the major product (25%), indicating that in the primary step 114 reacts with 115 to afford disaccharide 116 followed by desily-lation and coupling with the third component 113.

5.2
Enzyme-Mediated Oligosaccharide Synthesis

As discussed above, complex blocking/deblocking strategies as well as cumber-some workup procedures which often accompany classical glycosidation methods may be circumvented by enzyme-catalyzed oligosaccharide synthesis. Chemists have pursued three different strategies for in vitro enzymatic oligosaccharide synthesis [175, 227]. One approach uses glycosidases or glycosyl hydrolases as catalysts in transglycosylations, thus reversing the normal mode of action of these enzymes [228]; however, this method results in a mixture of regioisomers. A second approach requires enzymes of the non-Leloir pathway and sugar 1-phos-phates as glycosyl donors [229]. But by far the most widely applied method relies on glycosyltransferases from the Leloir pathway (Scheme 40) [230]. These enzym-es act by transferring carbohydrates from nucleoside diphosphate sugars 117 [231] to an acceptor 118 to afford glycoside 119 and nucleoside diphosphate 120.

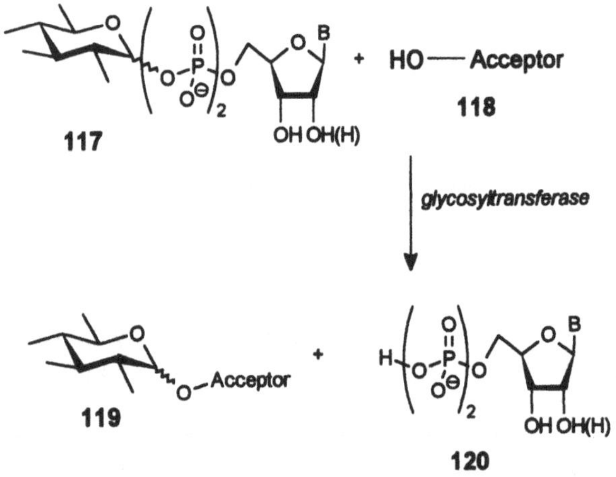

Scheme 40. Enzyme-catalyzed glycosidation

In contrast to glycosylhydrolases, glycosyltransferases usually afford a single product with a high degree of regio- and stereoselectivity.

For glycosyltransferase-mediated oligosaccharide synthesis, only a limited number of enzymes are currently available, of which galactosyl-, sialyl-, fucosyl-, N-acetylglucosaminyl- and mannosyltransferases are the most important examples. However, it should be noted that glycosyltransferases for mono- or multideoxygenated sugars have not been isolated, thus they have yet to be employed in enzyme-catalyzed oligosaccharide syntheses [232]. Furthermore, this strategy is hampered by the limited availability of nucleoside diphosphate-activated deoxygenated sugars. In particular the 2-deoxy derivatives are extremly difficult to handle, as they are highly sensitive to hydrolysis and 1,2-elimination [233]. Both chemical as well as enzymatic methods have been pursued to obtain sufficient amounts of these activated sugars. In this section we shall focus on the use of glycosyltransferases for the synthesis of deoxysugars and on how the requisite deoxygenated donor substrates can be provided with current techniques.

5.2.1
Synthesis of Nucleoside Diphosphate Sugars

5.2.1.1
Chemical Preparation

Several laboratories have reported the chemical synthesis of sugar nucleotides 117 that contain deoxygenated and otherwise modified sugars. All these syntheses have a biomimetic feature in that commonly a glycosylphosphate 121 is used as the starting point. This compound is then coupled with a nucleoside monophosphate (NMP) 122 usually activated as an amidate, preferentially a morpholidate, thereby following or slightly modifying Moffat's original procedure (Scheme 41) [234]. In some cases, phosphinothioic anhydrides served as

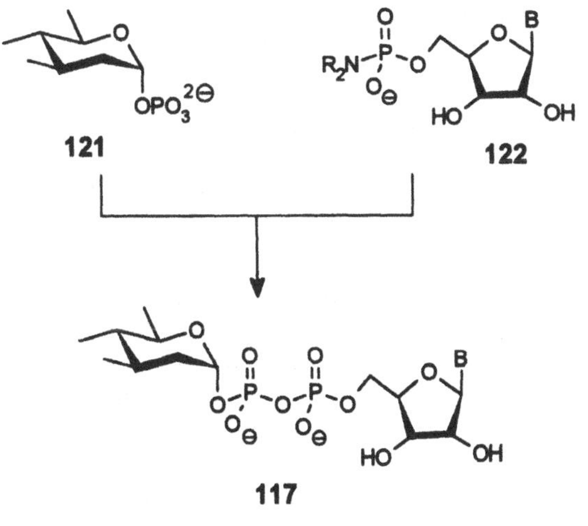

Scheme 41. Chemical strategy for the preparation of nucleoside diphosphate sugars

activated nucleoside monophosphates [235]. Alternatively, a reversed coupling mode of both fragments has been achieved by use of a glycosylphosphoric imidazolide and UMP [236].

A number of strategies for the synthesis of glycosyl 1-phosphates have been devised (Scheme 42). Among other methods [237], phosphorylation of protected sugars with diphenyl- [238] or 1,2-phenylene [239] phosphorochloridate to yield glycosyl-1-phosphates **123a–c**, which are further transformed by hydrogenolysis into **121**, has found wide acceptance. Recently, phosphoramidites have been employed to give the corresponding phosphite intermediates **124a–c** [240, 241] and related structures [242] which are further oxidized to the glycosyl phosphates **121**.

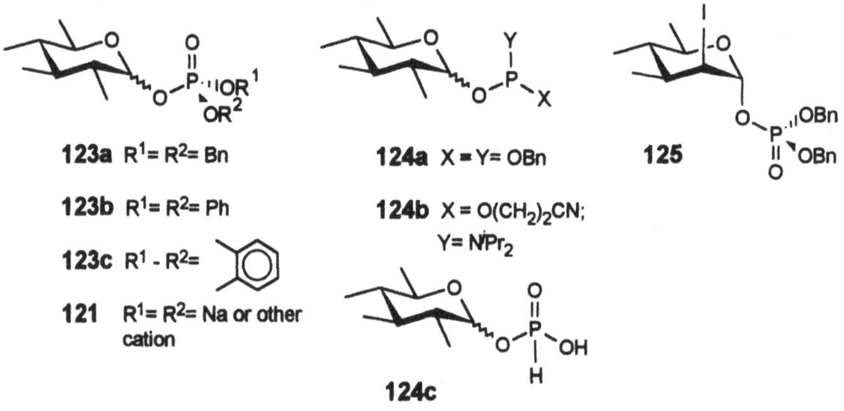

Scheme 42. Strategies for the synthesis of glycosyl 1-phosphates

Trichloroacetimidates [170c, 243], glycosyl bromides [235c, 244a] and allyl glycosides [244b] have also been utilized for the synthesis of glycosyl phosphates **123** and glycosyl phosphites **124**, respectively. When glycosyl chlorides are employed, silver dibenzyl [245] phosphate is usually needed as a phosphorylating reagent. Although deoxygenation at any position, in particular at C-2, of glycosyl 1-phosphates leads to greater instability, all of the above mentioned methods are applicable to 3-,4-or 6-deoxygenated sugar phosphates [240d, 246]. However, they usually fail for the synthesis of 2-deoxyglycosyl 1-phosphates!

Nevertheless, a few successful attempts towards the chemical synthesis of 2-deoxy-glycosyl 1-phosphates have been described in the literature. Glycosidation with 2-deoxyglycosyl chlorides in the presence of silver dibenzyl phosphate or iodonium-promoted addition of dibenzyl phosphate to glycals either affords the desired 2-deoxyglycosyl phosphates in low overall yield or gives the 2-iodo-2-deoxy derivatives **125** which are too labile for purification [247]. In fact, use of S-(2-deoxyglycosyl) phosphorodithioates [248] as glycosyl donors in iodonium cation-promoted glycosidations with dibenzyl phosphate is a very promising approach towards this class of glycosyl phosphates [247]. However, removal of the benzyl groups on phosphorous and coupling with an activated nucleoside is equally troublesome and has only successfully been achieved in a few cases.

For the important 6-deoxy nucleotide sugar GDP-fucose, various synthetic and a few enzymatic approaches have been devised (for a review see [227]). Hindsgaul and coworkers were the first to demonstate that nucleoside diphosphates of 2-deoxyhexoses could be prepared by chemical means (Scheme 43) [249a]. Their synthetic strategy avoided the use of acidic reaction conditions, as 1-O-phosphorylated 2-deoxyhexoses easily undergo elimination to afford glycals [233]. Deprotonation of the anomeric hydroxy group in pyranose 126 followed by phosphorylation [249b] yielded the corresponding phosphotriester, which was directly hydrogenated under weakly basic conditions to give glycosyl phosphate 127. De-O-acetylation and conversion into the sodium salt followed by coupling with uridine 5′-monophosphate morpholidate 128 afforded the very labile UDP-2-deoxy-galactose 129. Using ion-exchange chromatography, 80% purity was reached.

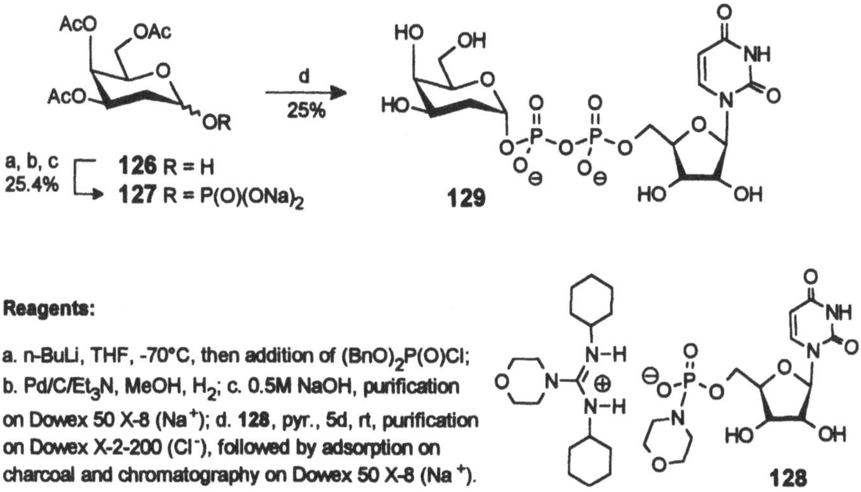

Reagents:

a. n-BuLi, THF, -70°C, then addition of (BnO)$_2$P(O)Cl;
b. Pd/C/Et$_3$N, MeOH, H$_2$; c. 0.5M NaOH, purification on Dowex 50 X-8 (Na$^+$); d. 128, pyr., 5d, rt, purification on Dowex X-2-200 (Cl$^-$), followed by adsorption on charcoal and chromatography on Dowex 50 X-8 (Na$^+$).

Scheme 43. First chemical synthesis of a NDP-2-deoxysugar

Recently, Schmidt and coworkers presented an elegant synthesis of the activated 6-deoxy-α-D-*ribo*-3-hexulose 130 (Scheme 44) [23]. Again the morpholidate methodology [234] was applied for the construction of the diphosphate unit. The stability of 130 is further decreased by the carbonyl group at C-3. The starting point of the synthesis was the ulose 131, which was converted into the pyranose 132 with an an exocyclic olefinic double bond at C-3 as a protective group. After formation of glycosyl phosphate 133, preparation of 130 was achieved by two alternative routes, either liberating the keto group from 133 or from 134 under mild ozonolytic reaction conditions.

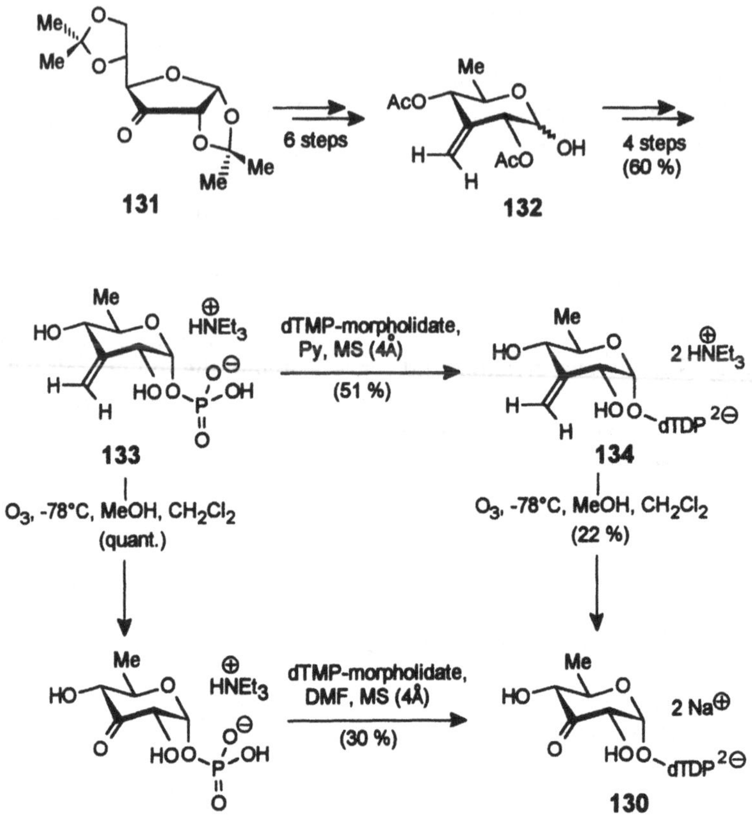

Scheme 44. Chemical synthesis of dTDP-6-deoxy-D-3-hexulose (130)

5.2.1.2
Enzymatic Preparation

In principle, there are two strategies for making nucleoside diphosphate sugars available by enzymatic means: (1) by in situ generation of activated sugars, thereby avoiding isolation steps [227a, 250], and (2) by large-scale preparation and isolation of nucleotide sugars [251]. Here, pyrophosphorylases, which catalyze the formation of nucleoside diphosphate sugars from sugar 1-phosphates and nucleoside triphosphates, have played a key role. However, enzyme-mediated synthesis of deoxygenated glycosyl phosphates and the corresponding nucleotide sugars is hampered by availability of the biocatalysts. Most kinases, pyrophosphorylases, glycosyltransferases and other enzymes from the metabolism of deoxygenated carbohydrates have not been described or isolated so far. Therefore, enzymes that usually accept conventional sugars as substrates have been employed for processing their deoxygenated counterparts, thus exploiting an inherent lack of substrate specificity. Indeed, in some cases this strategy turned out to be practical.

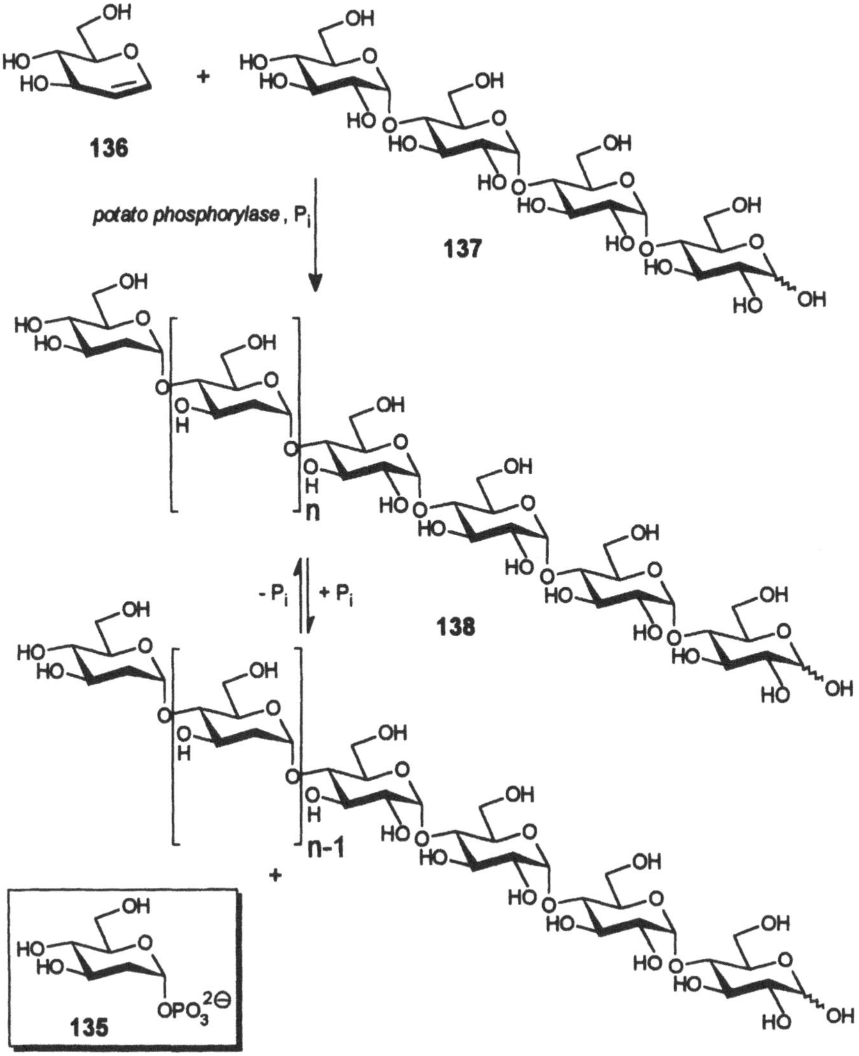

Scheme 45. Potato phosphorylase-catalyzed synthesis of 2-desoxy-D-glucose 1-phosphate 135

Potato phosphorylase has successfully been employed for the enzymatic synthesis of 2-deoxyglucosyl 1-phosphate **135** (Scheme 45) [252]. D-glucal **136** was transferred onto maltotetrose **137** in the presence of the phosphorylase and catalytic amounts of phosphate to afford a modified maltooligosaccharide **138** that contained 2-deoxy-D-arabino-hexopyranosyl residues [253]. Formation of 2-deoxyglucosyl phosphate **135** can be explained by reverse phosphorolysis of the oligosaccharide **138**. When soluble starch was employed as primer and an equimolar amount of phosphate was added, 2-deoxyglucosyl phosphate **135** was obtained in 50% isolated yield on a preparative scale.

However, the first enzyme-catalyzed synthesis of 2-deoxy-α-D-arabino-hexopyranosyl phosphate **135** was reported by Percival and Withers (Scheme 46)

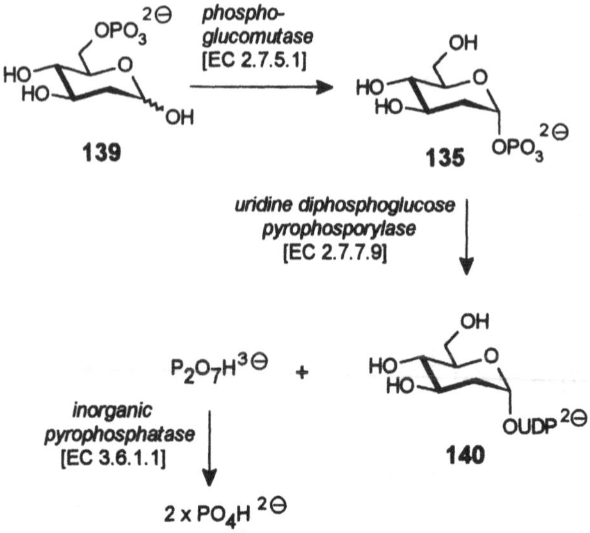

Scheme 46. Enzymatic synthesis of UDP-2-deoxy-D-glucose **140**

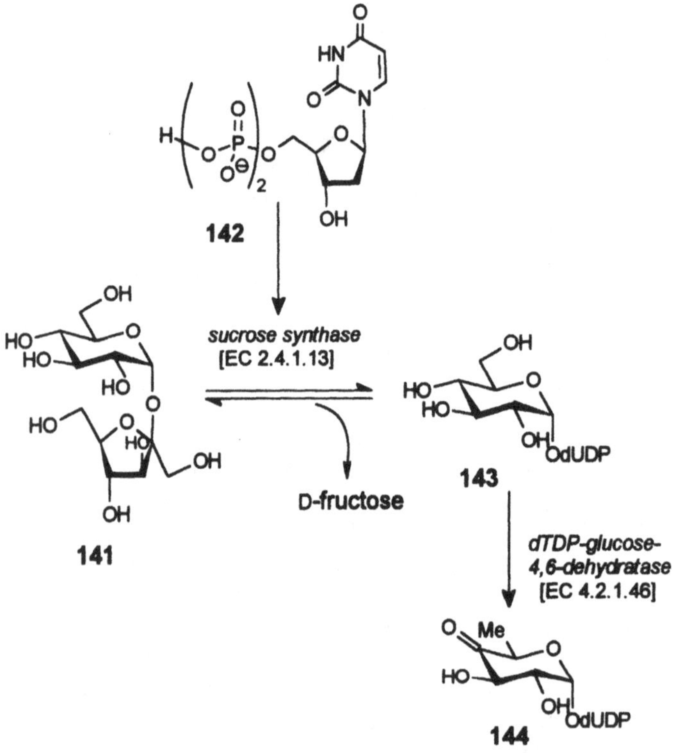

Scheme 47. Enzymatic synthesis of dUDP-6-deoxy-4-hexulose **144** from sucrose **141**

[32]. The starting point of their preparation was 2-deoxy-D-glucose-6-phosphate **139**, which can be obtained from D-glucose by hexokinase-mediated phosphorylation at C-6. The authors employed phosphoglucomutase to generate 2-deoxy-glycosyl 1-phosphate **135** in situ, which was directly transformed into UDP-2-deoxy-D-glucose **140** by uridine diphosphoglucose pyrophosphorylase. In order to pull formation of **140** to completion, inorganic pyrophosphatase was added thus removing pyrophosphate from the equilibrium. The reaction can be reversed to glycosyl 1-phosphate **135** by the same pyrophosphorylase in the presence of a large excess of pyrophosphate.

An interesting new enzymatic approach to nucleotide sugars has been described by Elling and coworkers [254]. They employed rice sucrose synthase [174], which catalyzes cleavage of sucrose **141**, with nucleoside diphosphates, e.g., 2'-deoxyuridine-5'-diphosphate **142**, to afford the corresponding activated sugar dUDP-D-glucose **143** along with D-fructose (Scheme 47). In addition, in situ transformation of **143** into dUDP-6-deoxy-α-D-*xylo*-4-hexulose **144** was performed in 49% isolated yield by adding dTDP-glucose 4,6-dehydratase (see Sects. 2.1.1 and 3.1.2) to the reaction mixture. As an extension of this work, Elling and Klaffke showed [173] that wild-type dTDP-glucose 4,6-dehydratase not only accepts the natural substrate **145a** but can also successfully be employed to transform the dTDP-sugars **145b, c** into the corresponding 4-keto-6-deoxy derivatives **146b, c** (Scheme 48).

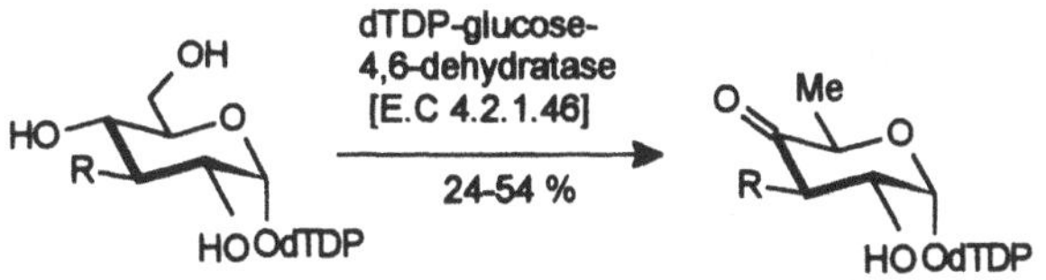

145a R= OH
145b R= H
145c R= N$_3$

146a R= OH
146b R= H
146c R= N$_3$

Scheme 48. Substrate specificity of dTDP-glucose-4,6-deydratase

5.2.2
Enzymatic Glycosidation

Several laboratories have reported chemical syntheses of sugar nucleotides that contain deoxygenated and otherwise modified sugars. Due to the current unavailability of glycosyltransferases from microbial sources that transfer deoxygenated sugars (see Sect. 2.2), these modified donor substrates have mainly been transferred by galactosyl transferase (GalT). As part of a project to develop specific glycosyltransferase inhibitors, the substrate specificity of fucosyl-(FucT) and N-acetylglucosaminyl-I-transferases (GnT-I) towards deoxygenated glycosyl acceptors has also been tested. In many cases it was shown that these enzymes indeed accept the glycosyl donors [235c, 246c, 255] and acceptors. However, the investigations further disclosed that glycosyltransferases may act with a high degree of substrate specificity and may show a low transfer rate for many deoxygenated sugar analogues and otherwise modified derivatives. Two variations of the glycosyltransferase-mediated construction of deoxygenated oligosaccharides have been pursued: Either suitable glycosyl donors, which were obtained by chemical or enzymatic synthesis, were employed in equimolar amounts with the corresponding NDP-derivatives or they were used as intermediates from enzymatic in situ generation.

Fucosyl transferases (FucT) play an important role in the biosynthesis of those oligosaccharides which determine blood group and which are active parts of cell surface and tumor antigens. Guanosine 5'-diphospho-β-L-fucose (GDP-Fuc) 147, the donor substrate for fucosyl transferases, has been prepared chemically [235c, 256] as well as enzymatically [257]. Interestingly, non-natural GDP-fucose derivatives such as GDP-3-deoxy-fucose 148 and GDP-L-arabinose 149 can be transferred onto suitable substrates of the Lewis[a] family, affording the corresponding trisaccharides 150 (Table 2) [235c].

Along these lines, Hindsgaul and coworkers also investigated the acceptor substrate specificity of a cloned α-(1→2) fucosyltransferase using structural analogues of octyl β-D-galactopyranoside 151 [258]. This transferase is one of two that is responsible for assembling the 0 blood group antigen using GDP-fucose 147 (see Table 2) as glycosyl donor. The analogues employed differed from the natural acceptor substrate by having replaced the hydroxy groups at C-3, C-4 and C-6 individually with deoxy 152–154, fluoro-, acetamido- or amino-functionalities. The substrate specificities for the deoxygenated analogues 152–154 was in line with those found for the other modified substrates prepared. The results are briefly summarized in Table 3. Deoxygenation at C-6 gave analogue

Table 2. Donorsubstrate specificity of Lewis α-(1→4)fucosyltransferase (Le-FucT)

	R	R[1]	Relative rate (%)
147	Me	OH (native)	100
148	Me	H	2.3
149	H	OH	5.9

Table 3. Acceptorsubstrate specificity of α-(1→2)fucosyltransferase

		Relative acceptor activity (%)
151	$R^1=R^2=R^3=$ OH (native)	100
152	$R^1=R^2=$ OH; $R^3=$H	64
153	$R^1=R^3=$ OH; $R^2=$ H	< 0.5 (K_i= 36.8 mM)
154	$R^1=$H; $R^2=R^3=$ OH	< 0.5 (K_i= 12.8 mM)

Table 4. Substrate specificity of deoxygenated acceptors for α-$(1 \to 3/4)$fucosyltransferase

		K_m (µM)	V_{max} (pmol/min µl)
155	R^1-R^5= OH	640	1.8
156a	R^1=H; R^2-R^3= OH	400	1.6
156b	R^2=H; R^1, R^3-R^5= OH	9.0 (K_i)	
156c	R^3=H; R^1=R^2=R^4=R^5= OH	730	3.2
156d	R^4=H; R^1-R^3, R^5= OH	850	1.6
156e	R^5=H; R^1-R^4= OH		

152, which showed pronounced substrate activity. The 3-deoxy pyranoside **154** is an inhibitor whereas the 4-deoxy analogue **153** is both a weak inhibitor and a substrate.

Another FucT studied in this context is Lewis $\alpha(1 \to 3/4)$fucosyltransferase, readily available from human milk. It transfers fucose from GDP-fucose **147** to both β-D-Gal $(1 \to 3)$-β-D-Glc NAc **155** and the corresponding $\beta(1 \to 4)$-linked analogue of **155** (Table 4). In an extension of their work, the Hindsgaul group prepared five different acceptor analogues of **155**, each of which had lost a different hydroxy group to hydrogen [259]. Disaccharides **156a – e** were kinetically evaluated as substrates as well as inhibitors of FucT. As listed in Table 4, deoxygenation at C-6 results in a better ligand **156a** for the enzyme, while removal of the 4-hydroxy group affords a disaccharide **156b** that is not a substrate but a very weak competitive inhibitor. Furthermore, deoxygenation at either C-2' or C-4' (**156c, d**) results in substrates whose activities were similar to the native substrate **155**. Only the C-6' deoxygenated analogue **156e** was totally inactive, both as acceptor and inhibitor, indicating that the 6'-OH group in **155** is essential for binding.

Galactosyltransferases have intensively been studied in terms of donor/acceptor substrate specificity [260]. Hashimoto and coworkers prepared two 6-modified UDP-D-galactoses **158** and **159** and investigated the bovine $\beta(1 \to 4)$ galactosyltransferase-mediated transfer to N-acetyl-D-glucosamine or its methyl glycoside (Table 5) [255]. Comparison with the native glycosyl donor UDP-D-galactose **157** revealed a highly reduced transfer rate. Nevertheless, the 6-modified N-acetyllactosamines **160b** (R^1 = H) and **160c** (R^1 = F) and other oligosaccharides were synthesized enzymatically in 30%–59% yields.

By following Whitesides' original concept [261], Thiem and Wiemann were the first to transfer an in situ-generated deoxy analogue of the natural donor

Table 5. Donorsubstrate specificity of β-(1→4)galactosyltransferase

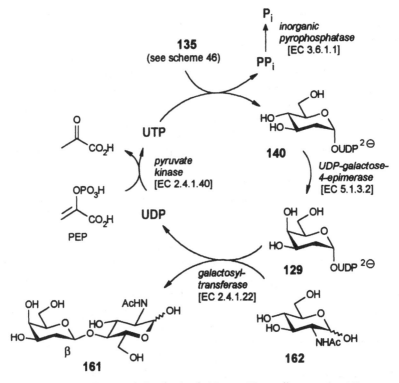

	R	R¹	Relative rate (%)
157	Me	OH (native)	100
158	H	H	1.3
159	Me	F	0.2

substrate which onto an acceptor by commercial β(1→4)-D-galactosyltransferase (GalT). They prepared 2′-deoxy-N-acetyllactosamine **161** by coupling UDP-2-deoxy-D-galactose **129** and N-acetylglucosamine **162** in the presence of GalT (Scheme 49) [262]. This example is particularly noteworthy, as here the glycosyltransferase converts a 2-deoxygenated substrate analogue into a β-glycoside with high efficiency and without the participation of a neighboring group (vide supra). In their strategy the labile 2-deoxy-1-phosphate **135** as well as the UDP-derivatives **129** and **140** are formed as intermediates of a cascade of enzyme reactions. Thus, **129** was prepared by epimerization of UDP-2-deoxy-D-glucose **140** in the presence of UDP-galactose-4-epimerase. Compound **140** was obtained from

Scheme 49. Enzymatic Synthesis of 2′-Deoxy-N-acetyllactosamine **161**

2-deoxy-D-glucose via glycosyl phosphate **135** according to Wither's work (see Scheme 46). To minimize inhibition effects and costs, uridine 5'-triphosphate (UTP) was constantly regenerated by use of phosphoenol pyruvate (PEP) and pyruvate kinase, thus closing the catalytic cycle.

It was further demonstrated that the kinetic parameters of UDP-2-deoxy-galactose **162** are almost identical to those of the natural substrate UDP-galactose. These results were fully confirmed by others [249a] when UDP-2-deoxy-D-galactose **129** was employed in an equimolar ratio (see also Scheme 43).

In a variant of the synthesis of 2'-deoxy-N-acetyllactosamine **161**, Wong and coworkers introduced galactokinase (GK) and galactose 1-phosphate uridyltransferase (Gal-1-P UT) as tools for the preparation UDP-2-deoxygalactose **129** (Scheme 50) [263]. The multienzyme synthesis starts from 2-deoxy-D-galactose **163**. By this alteration, a more direct enzymatic route to glycosylphosphate **164** is achieved, thereby avoiding an epimerization step. In addition, here UDP-glucose **165** can serve as a UMP transfer reagent which is regenerated from glucose 1-phosphate **166**.

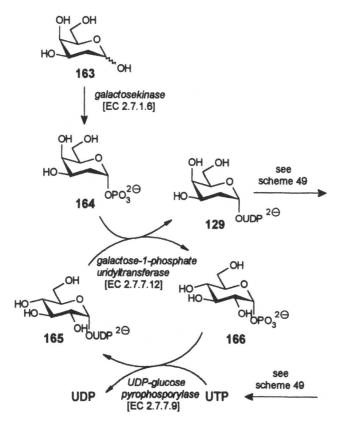

Scheme 50. Modified enzymatic route to UDP-2-desoxy-D-galactose **129**

6
Outlook

The past few years have seen a tremendous surge of interest in enzyme-catalyzed synthesis. Nonetheless, assembly of deoxygenated carbohydrates by enzymes is still hampered by the limited availability of both the corresponding glyco-syltransferases and the activated nucleoside diphosphate sugars. Preliminary experiments showed that carbohydrate moieties are usually built up one by one, and reaction is carried out by a glycosyltransferase that forms a specific linkage. However, these studies were conducted exclusively with glycosyltransferases, such as galactosyl- and fucosyltransferases, which transfer conventional sugars. In a few examples, deoxygenated monosaccharides have been used as non-natural acceptor or donor substrates (see Sect. 5.2.2). To some extent glycosidases may hold promise to overcome these obstacles [264]; however, success in utilizing the enzymes of the Leloir pathway will ultimately come from glycosyltransferases from microbial sources which are responsible for transfer of deoxygenated and otherwise modified sugars. Molecular biology, in connection with biosynthetic studies, is and will continue to be the central tool to obtain sufficient amounts of these biocatalysts, by discovering and cloning the corresponding genes and heterologously overexpressed them, e.g., in *E. coli*. Enzyme-mediated oligo-saccharide synthesis, and thereby biosynthetic studies on deoxysugars, will clearly benefit from the possibility of applying nonactivated putative inter-mediates. The successful incorporation of L-rhodinose into landomycin A (**12**) has paved the way for such approaches, since it now appears to be very likely that a general short activation pathway exists (see Sect. 2.3.3). In the future, these incorporation experiments may also be performed with nonlabeled, i.e., non-deuterated, sugars, since sophisticated NMR methods allow differentiation in a natural product between the original H/D ratio of the source (the added sugar) and that of the biosynthetically modified sugar building block. Since natural organisms always clearly prefer H over D (kinetic isotope effect), the H/D ratio is always increased if an original proton was replaced by another one during the biosynthetic cascade. These effects of such site-specific natural isotope fractio-nation (SNIF)-NMR experiments can be measured if the compounds are availa-bel in sufficient amounts [265]. The most famous example in this context was the differentiation between glucose in wine which arose from natural biosynthesis (photosynthesis) in the plant and sugar which was added later to "improve" the wine's quality [266].

Enzymes involved in the short activation pathway, probably a kinase plus a pyrophosphorylase, should exhibit broad substrate flexibility. Such enzymes will be powerful tools for constructing activated deoxysugars which are needed when utilizing glycosyltransferases in enzymatic synthesis.

Combinatorial biosynthetic approaches may soon also be extended to glyco-syltransferases. For example, transformation of glycosyltransferase encoding genes along with the genes controlling the biosynthesis of the required activated sugar building blocks may not be sufficient to obtain glycosylation of non-natural aglyca. Thus, genes responsible for proteins that support or even control glycosyl transfer of an activated sugar to an alcohol may be very useful by helping to increase substrate flexibility (see Sect. 2.2.3). Regarding precursor-directed biosynthetic approaches [265], the genetically engineered micro-organisms with heterologously expressing genes of the short activation pathway and those genes supporting glycosyl transfer (glycosyltransferases, etc.) may be advantageous. Such an engineered organism need only to be fed with an arsenal of deoxysugars in order to produce a wide spectrum of glycosylated seminatural products.

It is not only enzyme-mediated oligosaccharide synthesis that will soon help to bring glycobiology onto the center stage of biochemistry. In the future, chemical oligosaccharide synthesis will undergo a dramatic change when methodologies for assembling these biomolecules on ploymer supports are developed [267]. Preliminary studies [268] indicate that methods to reach this goal, although still extremely uneven and problematic compared to solid phase petide or oligonucleotide synthesis, have been found. Oligosaccharide assembly on polymer supports strategy has two very important potential advantages over classical solution chemistry: (1) reduction of the enormous costs of purification after each glycosidation step and (2) the establishment of combinatorial carbohydrate chemistry, which will be able to provide libraries of oligosaccharides for the pharmaceutical industry [269]. As larger numbers of oligosaccharides become available, their biological functions and therapeutic potential can be better studied and understood.

Acknowledgement. In view of a speculative and risky project, we thank the Deutsche Forschungsgemeinschaft for generously supporting our close collaboration with the grants Ki 397/2-1, 2-2 and 2-3; Ro 676/5-1 and 5-2; and Be 1389/3-1.

7
References

1. Kennedy JF, White CA (1983) Bioactive Carbohydrates in Chemistry, Biochemistry, and Biology. Ellis Horwood Publishers, Chichester
2. Haslam E (1993) Shikimic Acid, Metabolism and Metabolites. John Wiley and Sons, Chichester
3. Stryer L (1988) Biochemistry. W. H. Freeman and Co., New York
4. Metzler DE (1977) Biochemistry – The Chemical Reactions of Living Cells. Academic Press, New York
5. Gottschalk G (1986) Bacterial Metabolism. Springer Series in Microbiology. Springer-Verlag, Berlin
6. Voet D, Voet JG (1994) Biochemie. In: A. Maelicke; W. Müller-Esterl (eds) VCH, Weinheim
7. Lowe JB (1994) Carbohydrate Recognition in Cell-Cell Interaction. In: Fukuda M, Hindsgaul O (eds) Molecular Glycobiology. Oxford University Press, Oxford, p 163
8. Varki A (1994) Proc Natl Acad Sci USA 91:7390
9. Ichikawa Y, Lin Y-C, Dumas D-P, Shen G-J, Garcia-Junceda E, Williams MA, Bayer R, Ketcham C, Walker LE, Paulson JC, Wong C-H (1992) J Am Chem Soc 114:9283
10. Danishefsky SJ, Bilodeau MT (1996) Angew Chem 108:1482; Angew Chem Int Ed Engl 35:1380
11. Rouhi AM (1995) C & EN 1995:31
12. Bilodeau MT, Park TK, Hu S, Randolph JT, Danishefsky SJ (1995) J Am Chem Soc 117:7840
13. Gabius HJ (1988) Lectins and Glucoconjugates in Oncology. In: HJ Gabius, GA Nagel (eds) Springer, New York

14. Gabius HJ (1993) Biochim Biophys Acta 1071:1
15. Stüben J, Bertram B, Wießler M (1995) Int J Oncol 7:225
16. Pohl J, Bertram B, Hilgard P, Nowrousian MR, Stüben J, Wießler M (1995) Cancer Chemother Pharmacol 35:364
17. Wyss DF, Choi JS, Li J, Knoppers MH, Willis KJ, Arulanandam ARN, Smolyar A, Reinherz EL, Wagner G (1995) Science 269:1273
18. Lagunas R, Moreno E (1992) Yeast 8:107
19. Vertesy L, Fehlhaber H-W, Schulz A (1994) Angew Chem 106:1936
20. Ohta T, Hashimoto E, Hasegawa M (1992) J Antibiot 45:1167
21. Liu H-W, Thorson JS (1994) Annu Rev Microbiol 48:223
22. van der Donk WA, Stubbe J, Gerfen GJ, Bellew BF, Griffin RG (1995) J Am Chem Soc 117:8908
23. Müller T, Schmidt RR (1995) Angew Chem 107:1467; Angew Chem Int Ed Engl 34:1328
24. Vara J, Lewandowska-Skarbek M, Wang Y-G, Donadio S, Hutchinson CR (1989) J Bacteriol 171:5872
25. Rinehart KL, Snyder WC, Staley AL, Lau RCM (1992) Biosynthetic Studies on Antibiotics. In: Petroski RJ, McCormick SP (eds) Secondary-Metabolite Biosynthesis and Metabolism. Plenum Press, New York, p 41
26. Brockmann H, Waehnelt T (1963) Naturwiss 50:43
27. Stevens C-L, Blumbergs P, Wood DL (1964) J Am Chem Soc 86:3592
28. Ichikawa Y, Sim MM, Wong CH (1992) J Org Chem 57:2943
29. Liu L-D, Liu H-W (1989) Tetrahedron Lett 30:35
30. Srivastava G, Hindsgaul O, Palcic MP (1993) Carbohydr Res 245:137
31. Stiller R, Thiem J (1992) Liebigs Ann Chem 1992:467
32. Percival MD, Withers SG (1988) Can J Chem 66:1970
33. Kornfeld S, Glaser L (1961) Federation Proc 20:84
34. Kornfeld S, Glaser L (1960) Biochim Biophys Acta 42:548
35. Ginsburg V (1961) J Biol Chem 236:2389
36. Gabriel O (1968) Carbohydr Res 6:111
37. Gabriel O (1973) Adv Chem Ser 117:387
38. Zarkowsky H, Glaser L (1969) J Biol Chem 244:4750
39. Snipes CE, Brillinger G-U, Sellers L, Mascaro L, Floss HG (1977) J Biol Chem 252:8113
40. a) Russel RN, Liu H-W (1991) J Am Chem Soc 113:7777; b) Russel RN, Thorson JS, Liu L-D, Liu H-W (1992) J Biol Chem 267:5868
41. Otsuka H, Mascaretti OA, Hurley LH, Floss HG (1980) J Am Chem Soc 102:6817
42. Snipes CE, Chang C-J, Floss HG (1979) J Am Chem Soc 101:701
43. Lee JJ, Lee JP, Keller PJ, Cottrell CE, Chang C-J, Zähner H, Floss HG (1986) J Antibiot 39:1123
44. Cornforth JW, Redmond JW, Eggerer H, Buckel W, Gutschow C (1970) Eur J Biochem 14:1
45. Lüthy J, Rétey J, Arigoni D (1969) Nature 221:1213
46. a) Oths PJ, Mayer RM, Floss HG (1990) Carbohydr Res 198:91; b) In this case the stereochemistry is probably a consequence of the mechanism. If this mechanism is superior to other possible mechanisms, then even if independant evolution would result in enzymes that operate with the same stereochemistry. Hence, stereochemical constancy is in this case not an indicator of a common ancestor, although DNA sequence comparisons, of course, support this conclusion (see also Hanson KR, Rose IA (1975) Accts Chem Res 8:1 for a general discussion)
47. Rubenstein PA, Strominger JL (1974) J Biol Chem 249:3776
48. Rubenstein PA, Strominger JL (1974) J Biol Chem 249:3782
49. Shih Y, Yang D-Y, Weigel TM, Liu H-W (1990) J Am Chem Soc 112:9652
50. Thorson JS, Oh E, Liu H-W (1992) J Am Chem Soc 114:6941
51. Miller VP, Liu H-W (1992) J Am Chem Soc 114:1880
52. Thorson JS, Lo SF, Liu H-W (1993) J Am Chem Soc 115:5827
53. Thorson JS, Lo SF, Liu H-W, Hutchinson CR (1993) J Am Chem Soc 115:6993
54. Pieper PA, Guo Z, Liu H-W (1995) J Am Chem Soc 117:5158
55. Gould SJ, Guo J (1992) J Am Chem Soc 114:10176

56. Wohlert S-E, Rohr J (1996) Unpublished Results
57. Wohlert S-E (1994) Diplomarbeit, Universität Göttingen 1994
58. Corcoran JW (1981) Biochemical Mechanisms in the Biosynthesis of the Erythromycins. In: JW Corcoran (eds) Antibiotics, Biosynthesis. Springer, Berlin, p 132
59. Rohr J, Zeeck A (1990) Biogenetic-Chemical Classification of Secondary Metabolites Produced by Fermentation. In: RK Finn, P Präve (eds) Biotechnology Focus 2. Hanser Publishers, Munich, Vienna, New York, p 251
60. a) Truscheit E, Frommer W, Junge B, Müller L, Schmidt DD, Wingender W (1981) Angew Chem Int Ed Engl. 20:744; b) Degwert U, van Hülst R, Pape H, Herrold RE, Beale JM, Keller PJ, Lee JP, Floss HG (1987) J Antibiot 40:855; c) The classification of aminosugars as deoxysugars is a somewhat misleading formalism in terms of biosynthetic mechanisms. Ring B of acarbose, although formally a 4,6-dideoxysugar, is biosynthetically clearly not related to other "real" 4-deoxysugars, i.e. it doesn't require a true 4-deoxygenation, but a transamination. For the biosynthesis of acarbose the nitrogen may very well be introduced through ring B as 4-amino-6-deoxyglucose (transamination product of 4-keto-6-deoxy-glucose) instead of as shown in Scheme 20a
61. Hardick DJ, Hutchinson DW, Trew SJ, Wellington EMH (1992) Tetrahedron 48:6285
62. Hardick DJ, Hutchinson DW (1993) Tetrahedron 49:6707
63. Imada A, Kintaka K, Nakao M, Shinagawa S (1982) J Antibiot 35:1400
64. Shinagawa S, Kasahara F, Yoshikazu W, Harada S, Asai M (1984) Tetrahedron 40:3465
65. Tsukamoto N, Fujii I, Ebizuka Y, Sankawa U (1992) J Antibiot 45:1286
66. Cooper R, Unger S (1986) J Org Chem 51:3942
67. Sakuda S, Isogai A, Matsumoto S, Suzuki A (1987) J Antibiot 40:296
68. Zhou Z-Y, Sakuda S, Yamada Y (1992) J Chem Soc Perkin Trans 1 1992:1649
69. Zhou Z-Y, Sakuda S, Kinoshita M, Yamada Y (1993) J Antibiot 46:1582
70. Pearce CJ, Rinehart JKL (1981) Bioynthesis of Aminocyclitol Antibiotics. In: J W Corcoran (eds) Antibiotics, Biosynthesis. Springer-Verlag, Berlin, p 74
71. a) Goda SK, Akhtar M (1992) J Antibiot 45:984; b) Yamauchi N, Kakinuma K (1992) J Antibiot 45:756; c) Yamauchi N, Kakinuma K (1992) J Antibiot 45:774
72. a) Piepersberg W (1995) Streptomycin and Related Aminoglycoside Antibiotics. In: L Vining; C Stuttard (eds) Biochemistry and Genetics of Antibiotic Biosynthesis. Butterworth-Heinemann, Stoneham, p 71; b) Retzlaff L, Mayer G, Beyer S, Ahlert J, Verseck S, Distler J, Piepersberg W (1993) Streptomycin Production in Streptomycetes: A Progress Report. In: RH Baltz; GD Hegemann; PL Skatrud (eds) Industrial Microorganisms: Basic and Applied Molecular Genetics. American Society of Microbiology, Washington, DC 20005, p 183; c) Heinzel P, Werbitzky O, Distler J, Piepersberg W (1988) Arch Microbiol 150:184
73. a) Kawai H, Hayakawa Y, Nakagawa M, Furihata K, Seto H, Otake N (1984) Tetrahedron Lett 25:1941 b) Hütter K, Baader E, Frobel K, Zeeck A (1986) J Antibiot 39:1191; c) Kind R, Hütter K, Zeeck A (1988), unpublished
74. a) Ohta K, Mizuta E, Okazaki H, Kishi T (1984) Chem Pharm Bull 32:4350; b) Imamura N, Kakinuma K, Ikekawa N, Tanaka H, Omura S (1981) Chem Pharm Bull 29:1788
75. a) Thiem J, Schneider G (1983) Angew Chem 95:54; b) Krishna NR, Miller DM, Sakai TT (1990) J Antibiot 43:1543; c) Nakano H, Ogawa H, Yamashita Y, Katahira R, Chiba S, Iwasaki T, Ashizawa T (1995) PCT Int Appl WO 95 06,054; Chem Abstr 123:8030p
76. Kupfer E, Neupert-Laves K, Dobler M, Keller-Schierlein H (1982) Helv Chim Acta 65:3
77. Krohn K, Rohr J (1997) Top Curr Chem 188:127–195
78. Saitoh K, Furumai T, Oki T, Nishida F, Harada K-I, Suzuki M (1995) J Antibiot 48:162
79. Kirschning A, Ries M, Dräger G (1996) Unpublished Results
80. Weber S, Zolke C, Rohr J, Beale JM (1994) J Org Chem 59:4211
81. Weißbach U (1996) Diplomarbeit, Universität Göttingen
82. Weißbach U, Rohr J (1996) Unpublished Results
83. Gerlitz M, Hammann P, Thiericke R, Rohr J (1992) J Org Chem 57:4030
84. Gerlitz M (1995) Erkenntnis- und anwendungsorientierte Untersuchungen zur Polyketid-Biosynthese. Dissertation, Universität Göttingen. Cuvillier Verlag, Göttingen

85. Gerlitz M, Rohr J (1995) Unpublished Results
86. Wagner C, Eckardt K, Ihn W, Schumann G, Stengel C, Fleck WF, Tresselt D (1991) J Basic Microbiol 31:223
87. Bartel PL, Connors NC, Strohl WR (1990) J General Microbiol 136:1877
88. Grimm A, Madduri K, Ali A, Hutchinson CR (1994) Gene 151:1
89. Weber JM, Leung JO, Swanson SJ, Idler KB, McAlpine JB (1991) Science 252:114
90. Baltz RH, Seno ET, Stonesifer J, Wild GM (1983) J Antibiot 36:131
91. Rohr J, Schönewolf M, Udvarnoki G, Eckardt K, Schumann G, Wagner C, Beale JM, Sorey SD (1993) J Org Chem 58:2547
92. Rohr J, Thiericke R (1992) Nat Prod Rep 9:103
93. Rohr J (1989) J Antibiot 42:1482
94. Rohr J (1990) Angew Chem Int Ed Engl 29:1051
95. Rohr J (1990) J Chem Soc Chem Commun 1990:113
96. Horii S, Kameda Y (1972) J Chem Soc Chem Commun 1972:747
97. Goeke K, Drepper A, Pape H (1996) J Antibiot 49:661
98. Toyokuni T, Jin W-Z, Rinehart JKL (1987) J Am Chem Soc 109:3481
99. Matselyukh B, Polishchuk L, Weißbach U, Wohlert S-E, Rohr J (1996) J Antibiot, In Preparation
100. Beninga C, Wohlert S-E, Rohr J (1996) Unpublished Results
101. Beninga C (1994) Diplomarbeit, Universität Göttingen
102. Piepersberg W (1994) Critical Rev Biotechnol 14:251
103. Piepersberg W (1994) BioEngineering 10:27
104. Rohr J (1995) Angew Chem Int Ed Engl 34:881
105. Tsoi CJ, Khosla C (1995) Chemistry & Biology 2:355
106. Katz L, Donadio S (1993) Annu Rev Microbiol 47:875
107. Hutchinson CR, Fujii I (1995) Annu Rev Microbiol 49:201
108. Rouhi AM (1995) C&EN 1995:9
109. Roessner CA, Scott AI (1996) Chemistry & Biology 3:325
110. Cundliffe E (1992) Antimicrob Agents Chemother 36:348
111. Decker H, Haag S, Udvarnoki G, Rohr J (1995) Angew Chem Int Ed Engl 34:1107
112. Decker H, Haag S (1995) J Bacteriol 177:6126
113. Decker H, Rohr J, Motamedi H, Zähner H, Hutchinson CR (1995) Gene 166:121
114. Decker H, Rohr J, Motamedi H, Hutchinson CR, Zähner, H (1995) Biotekhnologiya 1995:68
115. a) Wohlert SE, Decker H, Haag S, Rohr J (1996) Unpublished Results; b) The stereochemistry of the linked D-olivose in olivosyltetracenomycin requires an unlikely double inversion, as shown in Scheme 24. Alternatively, a glycosyltransferase in cosmid 16F4 may be proposed which is flexible regarding the sugar substrate. This flexibility may have been "caused" accidentally during the preparation of cosmid 16F4. This attractive alternative is currently under investigation
116. Rohr J, Beale JM, Floss HG (1989) J Antibiot 42:1151
117. Lau RCM (1990) Biosynthetic and Structural Studies of Neomycin and Berninamycin. Ph.D. Thesis. University of Illinois, Urbana-Champaign
118. Bekesi JG, Winzler RJ (1967) J Biol Chem 242:3873
119. Reitman ML, Trowbridge IS, Kornfeld S (1980) J Biol Chem 255:9900
120. Wohlert SE, Oelkers C, Ries M, Kirschning A, Rohr J (1997) J Chem Soc Chem Commun 1997:submitted
121. Kirschning A, Hary U, Ries M (1995) Tetrahedron 51:2297
122. a) Bechthold A, Sohng JK, Smith TM, Chu X, Floss, HG (1995) Mol Gen Genet 248:610; b) Westrich L, Faust B, Domann S, Bechthold A (1996) unpublished results; c) Sohng JK, August P, Yoo JC, Park S, Floss HG (1996) unpublished results; d) Tornus D, Floss HG (1996) 6th Conference on Genetics and Molecular Biology of Industrial Microorganisms, Bloomington, IN Abstract No P59
123. Reeves P (1993) TIG 9:17
124. Parsons FT, Preiss J (1978) J Biol Chem 253:7638
125. Parsons TF, Preiss J (1978) J Biol Chem 253:6197

126. Smith-White BJ (1992) J Mol Biol 34:449
127. Katsube T, Kazuta Y, Tanizawa K, Fukui T (1991) Biochemistry 30:8546
128. Lindquist L, Kaiser R, Reeves PR, Lindberg AA (1993) Eur J Biochem 211:763
129. Stockmann M, Piepersberg W (1992) FEMS Microbiol Lett 90:185
130. Decker H, Gaisser S, Pelzer S, Schneider P, Westrich L, Wohlleben W, Bechthold A (1996) FEMS-Microbiol Lett 141:195
131. a) Thorson JS, Liu H-W (1993) J Am Chem Soc 115:6998; b) Thorson JS, Liu H-W (1993) J Am Chem Soc 115:7539
132. Marumo, K, Lindqvist L, Verma N, Weintraub A, Reeves PR, Lindberg AA (1992) Eur J Biochem 204:539
133. Matsuhashi S (1966) J Biol Chem 241:4275
134. Matsuhashi S, Matsuhashi M, Strominger JL (1966) J Biol Chem 241:4267
135. Chang S, Buerr B, Serif G (1988) J Biol Chem 263:1693
136. Niemi J, Mäntsala P (1995) J Bacteriol 177:2942
137. Kanagasundaram V, Scopes RK (1992) J Bacteriol 174:1439
138. Loos H, Krämer R, Sahm H, Sprenger GA (1994) J Bacteriol 176:7688
139. Strohl W, Ye J, Dickens ML (1995) Biotecknologiya 45:7
140. Otten SL, Liu X, Ferguson J, Hutchinson CR (1995) J Bacteriol 177:6688
141. Hernandez C, Olano C, Mendez C, Salas JA (1993) Gene 134:139
142. Liu D, Haase AM, Lindquist L, Lindberg AA, Reeves P (1993). J Bacteriol 175:3408
143. Merson-Davies LA, Cundliff E (1994) Mol Microbiol 13:349
144. Gallo MA, Ward J, Hutchinson CR (1996) Microbiol 142:269
145. Peschke U, Schmidt H, Zhang H-Z, Piepersberg W (1995) Molecular Microbiol 16:1137
146. Linton KJ, Jarvis BW, Hutchinson CR (1995) Gene 153:33
147. Krügel H, Schumann G, Hänel F, Fiedler G (1993) Mol Gen Genet 241:193
148. a) Gräfe U (1992) Biochemie der Antibiotica. Spektrum Akademischer Verlag, Heidelberg, Berlin, New York; b) Zeeck A (1987) Schriftenreihe des Fonds der Chemischen Industrie 31:55
149. Hare DR, Wemmer DE, Chou SH, Drobny G, Reid BR (1983) J Mol Biol 171:319
150. Kennard O, Hunter WN (1991) Angew Chem 103:1280; Angew Chem Int Ed Engl 30:1254
151. a) Moore MH, Hunter, WN, Langlois B, d'Estaintot, Kennard O (1989) J Mol Biol 206:693; b) Wang AHJ, Ughetto G, Quigley GJ, Rich A (1987) Biochemistry 26: 1152; c) Williams LD, Egli M, Ughetto G, van der Marel GA, van Boom JH, Quigley GJ, Wang AHJ (1990) Biochemistry 29:2538
152. a) Hansen M, Hurley L (1995) J Am Chem Soc 117:2421; b) Sun D, Hansen M, Hurley L (1995) J Am Chem Soc 117:2430; c) Hansen M, Lee S-J, Cassady JM, Hurley LH (1996) J Am Chem Soc 118:5553
153. a) Zhang X, Patel DJ (1990) Biochemistry 29:9451; b) Searle MS, Bicknell W (1992) Eu J Biochem 205:45
154. Chen H, Patel DJ (1995) J Am Chem Soc 117:5901
155. a) Smith AL, Nicolaou KC (1996) J Med Chem 39:2103; b) Lee MD, Ellestad GA, Borders DB (1991) Acc Chem Res 24:235; c) Nicolaou KC, Dai W-M (1991) Angew Chem 103:1453; Angew Chem Int Ed Engl 30:1387
156. a) Walker S, Valentine KG, Kahne D (1990) J Am Chem Soc 112:6428; b) Walker S, Kahne D (1991) J Am Chem Soc 113:4716; c) Walker S, Murnick J, Kahne (1993) J Am Chem Soc 115:7954; d) Walker SL, Andreotti AH, Kahne D (1994) Tetrahedron 50: 1351
157. a) Aiyar J, Danishefsky SJ, Crothers DM (1992) J Am Chem Soc 114:7552; b) Nicolaou KC, Tsay SC, Suzuki T, Joyce GF (1992) J Am Chem Soc 114:7555
158. Gomez-Paloma L, Smith JA, Chazin WJ, Nicolaou, KC (1994) J Am Chem Soc 116:3697
159. Halcomb RL, Boyer SH, Danishefsky SJ (1992) Angew Chem 104:314; Angew Chem Int Ed Engl 31:338
160. Hawley RC, Kiessling LL, Schreiber SL (1989) Proc Natl Acad Sci USA 86:1105
161. Li T, Zeng Z, Estevez VA, Baldenius KU, Nicolaou KC, Joyce GF, (1994) J Am Chem Soc 116:3709

162. Calicheamicin Θ is a derivative of calicheamicin γ$_1^I$, which lacks the trisulfide trigger for stability reasons
163. Drak J, Iwasawa, N, Danishefsky SJ, Crothers D (1991) Proc Natl Acad Sci USA 88:7464
164. Ho SN, Boyer SH, Schreiber SL, Danishefsky SJ, Crabtree GR (1994) Proc Natl Acad Sci USA 91:9203
165. Nicolaou KC, Ajito K, Komatsu H, Smith BM, Li T, Egan MG, Gomez-Paloma L (1995) Angew Chem 107:614; Angew Chem Int Ed Engl 34:576
166. Nicolaou KC, Smith BM, Ajito K, Komatsu H, Gomez-Paloma L, Tor Y (1996) J Am Chem Soc 118:2303
167. Liu C, Smith BM, Ajito K, Komatsu H, Gomez-Paloma L, Li T, Theodorakis EA, Nicolaou KC, Vogt PK (1996) Proc Nat Acad Sci USA 93:940
168. Bifulco G, Galeone A, Gomez-Paloma L, Nicolaou KC, Chazin WJ (1996) J Am Chem Soc 118:8817
169. a) Sugiura Y. Uesawa Y, Takahashi Y, Kuwahara J, Golik J, Doyle TW (1989) Proc Natl Acad Sci USA 86:7672; b) Tsuji T, Morioka H, Takezawa M, Ando T, Murai A, Shibai H (1986) Agric Biol Chem 50:1697
170. Reviews: a) Paulsen H (1982) Angew Chem 94:184; Angew Chem Int Ed Engl 21:155; b) Paulsen H (1984) Chem Soc Rev 13:15; c) Schmidt RR (1986) Angew. Chem. 98:213; Angew Chem Int Ed Engl 25:212; d) Thiem J, Klaffke W (1990) Topics Current Chem. 54:285; e) Toshima K, Tatsuta K (1993) Chem Rev 93:1503; f) Lockhoff O (1992). In: Hagemann H, Klamann D (eds) Houben Weyl Methoden der Organischen Chemie E14a, part 3. Georg Thieme Verlag, Stuttgart, New York, p 621; g) Danishefsky SJ, Roberge JY (1995) Pure Appl Chem 67:1647; h) Boons G-J (1996) Tetrahedron 52:1095
171. a) Mootoo DR, Konradsson P, Udodong U, Fraser-Reid B (1989) J Am Chem Soc 111:8540; b) Konradsson P, Mootoo DR, McDevitt RE, Fraser-Reid B (1990) J Chem Soc Chem Commun 631; c) Fraser-Reid B, Wu Z, Udodong U, Ottosson H (1990) J Org Chem 55:6068
172. Randolph JT, McClure KF, Danishefsky SJ (1995) J Am Chem Soc 117:5712
173. a) Stein A, Kula M-R, Elling L, Verseck S, Klaffke W (1995) Angew Chem 107:1881; Angew Chem Int Ed Engl 34:1748; b) Naundorf A, Klaffke W (1996) Carbohydr Res 285:141
174. Elling L, Güldenberg B, Grothus M, Zervosen A, Peus M, Helfer A, Stein A, Adrian H, Kula M-R (1995) Biotechnol Appl Biochem 21:29
175. Bednarski MD, Simon ES (eds) (1991) In: Enzymes in Carbohydrate Chemistry, ACS Symposium Series 466
176. a) Martin A, Pais M, Monneret C (1983) J Chem Soc Chem Commun 306; b) Garegg PJ, Köpper S, Ossowski P, Thiem J (1986) J Carbohydr Chem 5:59; c) Brinkley RW (1990) J Carbohydr Chem 9:507
177. Although not all of the listed promoters have been applied to the synthesis of 2-deoxy glycosides it is reasonable to assume that they can also be used for this purpose
178. a) Wuts PGM, Bigelow SS (1983) J Org Chem 48:3489; b) Takeda K, Nakamura H, Ayabe A, Akiyama A, Harigaya Y, Mizuno Y (1994) Tetrahedron Lett 35:125; c) Takeda K, Kaji E, Nakamura H, Akiyama A, Konda Y, Mizuno Y, Takayanagi H, Harigaya Y (1996) Synthesis 341
179. Mukaiyama T, Nakatsuka T, Shoda S (1979) Chem Lett 487
180. Kihlberg JO, Leigh DA, Bundle DR (1990) J Org Chem 55:2860
181. a) Ferrier RJ, Hay RW, Vethaviyasar N (1973) Carbohydr Res 27:55; b) van Cleve JW (1979) Carbohydr Res 70:161; c) Garegg PJ, Henrichson C, Norberg T (1983) Carbohydr Res 116:162; d) Tsai TYR, Jin H, Wiesner K (1984) Can J Chem 62:1403
182. a) Sato S, Mori M, Ito Y, Ogawa T (1986) Carbohydr Res 155:C6; b) Sato S, Ito Y, Ogawa T (1988) Tetrahedron Lett 29:4759; c) Sakai K, Nakahara Y, Ogawa T (1990) Tetrahedron Lett 31:3035
183. a) Ito Y, Ogawa T (1988) Tetrahedron Lett 29:1061; b) Mori M, Ito Y, Uzawa J, Ogawa T (1990) Tetrahedron Lett 31:3191; c) Matsuzaki Y, Nunomura S, Ito Y, Sugimoto M, Nakahara Y, Ogawa T (1993) Carbohydr Res 242:C1; d) Fujita S, Numata M, Sugimoto M, Tomita K, Ogawa T (1994) Carbohydr Res 263:181

184. a) Nicolaou KC, Seitz SP, Papahatjis DP (1983) J Am Chem Soc 105:2430; b) Roush WR, Lin X, Straub JA (1991) J Org Chem 56:1649; c) Fukase K, Hasuoka A, Kusumoto S (1993) Tetrahedron Lett 34:2187

185. a) Veeneman GH, van Boom JH (1990) Tetrahedron Lett 31:275; b) Veeneman GH, van Leuwen SH, van Boom JH (1990) Tetrahedron Lett 31:1331; c) Konradsson P, Udodong UE, Fraser-Reid B (1990) Tetrahedron Lett 31:4313; d) Smid P, de Ruiter GA, van der Marel GA, Rombouts FM, van Boom JH (1991) J Carbohydr Chem 10:833; e) Ehara T, Kameyama A, Yamada Y, Ishida H, Kiso M, Hasegawa A (1996) Carbohydr Res 281:237

186. a) Lönn H (1985) Carbohydr Res 139:105; b) Lönn H (1987) J Carbohydr Chem 6: 301; c) Dasgupta F, Garegg PJ (1990) Carbohydr Res 202:225; d) Mereyala HB, Kulkarni VR, Ravi D, Sharma GVM, Rao BV, Reddy GB (1992) Tetrahedron 48:545

187. a) Sun L, Li P, Zhao K (1994) Tetrahedron Lett 35:7147; b) Fukase K, Kinoshita I, Kanoh T, Nakai Y, Hasuoka A, Kusumoto S (1996) Tetrahedron 52:3897

188. a) Nicolaou KC, Chucholowski A, Dolle RE, Randall JL (1984) J Chem Soc Chem Commun 1155; b) Kunz H, Sager W (1985) Helv Chim Acta 68:283

189. a) Nicolaou KC, Ladduwahetty T, Randall JL, Chucholowski A (1986) J Am Chem Soc 108:2466; b) Nicolaou KC, Hummel CW, Bockovich NJ, Wong C-H (1991) J Chem Soc Chem Commun 870

190. Hashimoto S, Hayashi M, Noyori R (1984) Tetrahedron Lett 25:1379

191. a) Kreuzer M, Thiem J (1986) Carbohydr Res 149:347; b) Jünnemann J, Lundt I, Thiem J (1991) Liebigs Ann Chem 759

192. a) Matsumoto T, Maeta H, Suzuki K, Tsuchihashi G (1988) Tetrahedron Lett 29:3567 and 3575; b) Matsumoto T, Katsuki M, Suzuki K (1989) Chem Lett 437

193. Suzuki K, Maeta H, Suzuki T, Matsumoto T (1989) Tetrahedron Lett 30:6879

194. a) Nicolaou KC, Caulfield TJ, Kataoka H, Stylianides NA (1990) J Am Chem Soc 112:3693; b) Nicolaou KC, Hummel CW, Iwabuchi Y (1992) J Am Chem Soc 114:3126

195. a) Mukaiyama T, Murai Y, Shoda S (1981) Chem Lett 431; b) Nicolaou KC, Dolle RE, Papahatjis DP, Randall JL (1984) J Am Chem Soc 106:4189; c) Nicolaou KC, Randall JL, Furst GT (1985) J Am Chem Soc 107:5556

196. Kobayashi S, Koide K, Ohno M (1990) Tetrahedron Lett 31:2435

197. a) Ogawa T, Beppu K, Nakabayashi S (1981) Carbohydr Res 93:C6; b) Mukaiyama T, Shimpuku T, Takashima T, Kobayashi S (1989) Chem Lett 145; c) Mukaiyama T, Takashima T, Katsurada M, Aizawa H (1991) Chem Lett 533; d) Mukaiyama T, Katsurada M, Shimpuku T, Takashima T (1991) Chem Lett 985; e) Petráková E, Glaudemans CPJ (1995) Carbohydr Res 268:35

198. a) Kimura Y, Suzuki M, Matsumoto T, Abe R, Terashima S (1984) Chem Lett 501; b) Jütten P, Scharf H-D, Raabe G (1991) J Org Chem 56:7144

199. Ley SV, Armstrong A, Diez-Martin D, Ford MJ, Grice P, Knight JG, Kolb HC, Madin A, Marby CA, Mukherjee S, Shaw AN, Slawin AMZ, Vile S, White AD, Williams DJ, Woods M (1991) J Chem Soc, Perkin Trans 1:667

200. Marra A, Gauffeny F, Sinaÿ P (1991) Tetrahedron 47:5149

201. a) Tietze L-F, Fische R, Guder H-J (1982) Tetrahedron Lett 23:4661; b) Priebe W, Grynkiewicz G, Neamati N (1991) Tetrahedron Lett 32:2079; c) Kolar C, Kneissl G, Knödler U, Dehmel K (1991) Carbohydr Res 209:89

202. a) Müller T, Schneider R, Schmidt RR (1994) Tetrahedron Lett 35:4763; b) Wu S-H, Shimazaki M, Lin C-C, Qiao L, Moree WJ, Weitz-Schmidt G, Wong C-H (1996) Angew Chem 108:106; Angew Chem Int Ed Engl 35:88; c) Hashimoto S, Sano A, Sakamoto H, Nakajima M, Yanagiya Y, Ikegami S (1995) Synlett 1271

203. a) Danishefsky SJ, Halcomb RL (1989) J Am Chem Soc 111:6661; b) McClure KF, Randolph JT, Ruggeri R, Danishefsky SJ (1993) Science 260:1307

204. a) Toshima K, Tatsuta K, Kinoshita M (1988) Bull Chem Soc Jpn 61:2369; b) Tu CJ, Lednicer D (1987) J Org Chem 52:5624; c) Bolitt V, Mioskowski C, Lee C-G, Falck JR (1990) J Org Chem 55:5812; d) Sabesan S, Neira S (1991) J Org Chem 56:5468

205. a) Lemieux RU, Morgan AR (1965) Can J Chem 43:2190; b) Thiem J, Karl H, Schwentner J (1978) Synthesis 696; c) Friesen RW, Danishefsky SJ (1990) J Am Chem Soc 111:6656

206. Tatsuta K, Fujimoto K, Kinoshita M, Umezawa S (1977) Carbohydr Res 54:85
207. Jaurand G, Beau J-M, Sinaÿ P (1981) J Chem Soc Chem Commun 572
208. Thiem J, Klaffke W (1989) J Org Chem 54:2006
209. a) Leblanc Y, Fitzsimmons BJ, Springer JP, Rokach J (1989) J Am Chem Soc 111:2995;
     b) Toepfer A, Schmidt RR (1993) Carbohydr Res 247:159
210. Capozzi G, Dios A, Franck RW, Geer A, Marzabadi C, Menichetti S, Nativi C, Tamarez M
     (1996) Angew Chem 108:805; Angew Chem Int Ed Engl 35:777
211. Wiesner K, Tsai TYR, Jin H (1985) Helv Chim Acta 68:300
212. Brinkley RW, Koholic DJ (1988) J Carbohydr Chem 7:487
213. Bock K, Pedersen M, Thiem J (1979) Carbohydr Res 73:85; b) Thiem J, Gerken M (1982)
     J Carbohydr Chem 1:229; c) Thiem J, Gerken M, Bock K (1985) Liebigs Ann Chem 462;
     d) Thiem J, Gerken M (1985) J Org Chem 50:954; e) Thiem J, Gerken M, Schöttmer B,
     Weigand J (1987) Carbohydr Res 164:327; f) Thiem J, Schöttmer B (1987) Angew Chem
     99:591; Angew Chem Int Ed Engl 26:555
214. Perez M, Beau J-M (1989) Tetrahedron Lett 30:75
215. Ito Y, Ogawa T (1987) Tetrahedron Lett 28:2723
216. a) Ramesh S, Kaila N, Grewal G, Franck RW (1990) J Org Chem 55:5; b) Grewal G, Kaila N,
     Franck RW (1992) J Org Chem 57:2084; c) Franck RW, Kaila N (1993) Carbohydr Res 239:71
217. Zuurmond HM, van der Klein PAM, van der Marel GA, van Boom JH (1993) Tetrahedron
     49:6501
218. Maharaj VJ, Senabe JV (1995) International Conference on Biological Challenges for
     Organic Chemistry, St Andrews, Scotland, Abstr P71
219. a) Toshima K, Mukaiyama S, Ishiyama T, Tatsuta K (1990) Tetrahedron Lett 31:3339 and
     6361; b) Toshima K, Mukaiyama S, Yoshida T, Tamai T, Tatsuta K (1991) Tetrahedron Lett
     32:6155; c) Toshima K, Nozaki Y, Mukaiyama S, Tatsuta K (1992) Tetrahedron Lett
     33:1491; d) Toshima K, Nozaki Y, Inokuchi H, Nakata M, Tatsuta K, Kinoshita M (1993)
     Tetrahedron Lett 34:1611
220. a) Nicolaou KC, Groneberg RD, Miyazaki T, Stylianides NA, Schulze TJ, Stahl W, (1990)
     J Am Chem Soc 112:8193; b) Groneberg RD, Miyazaki T, Stylianides NA, Schulze TJ,
     Stahl W, Schreiner EP, Suzuki T, Iwabuchi Y, Smith AL, Nicolaou KC (1993) J Am Chem
     Soc 115:7593
221. a) Halcomb RL, Boyer SH, Wittman M, Olson S, Denhardt D, Liu KKC, Danishefsky SJ
     (1995) J Am Chem Soc 117:5720; b) Hitchcock SA, Chu Moyer MM, Danishefsky SJ (1995)
     J Am Chem Soc 117:5750
222. a) Nicolaou KC, Hummel CW, Nakada M, Shibayama K, Pitsinos EN, Saimoto H,
     Mizuno Y, Baldenius K-U, Smith AL (1993) J Am Chem Soc 115:7625; b) Hitchcock SA,
     Boyer SH, Chu-Moyer MY, Olson SH, Danishefsky SJ (1994) Angew Chem 106:928;
     Angew Chem Int Ed Engl 33:858
223. a) Köpper S, Thiem J (1994) Carbohydr Res 260:219; b) Ley SV, Priepke HWM (1994)
     Angew Chem 106:2412; Angew Chem Int Ed Engl 33:2292; c) Yamada H, Harada T,
     Miyazaki H, Takahashi T (1994) Tetrahedron Lett 35:3979; d) Yamada H, Harada T,
     Takahashi T (1994) J Am Chem Soc 116:7919
224. Raghavan S, Kahne D (1993) J Am Chem Soc 115:1580
225. Bieber LW, DaSilva Filho AA, De Mello JF, De Lima OG, Do Nascimento MS, Veith HJ, Von
     der Saal W (1987) J Antibiot 40:1335
226. Suzuki K, Sulikowski GA, Friesen RW, Danishefsky SJ (1990) J Am Chem Soc 112:8895
227. a) Ichikawa Y, Look GC, Wong C-H (1992) Anal Biochem 202: 215; b) Wong C-H,
     Halcomb RL, Ichikawa Y, Kajimoto T (1995) Angew Chem 107:453 and 559; Angew Chem
     Int Ed Engl 34:412 and 511
228. Kornfeld R, Kornfeld S (1985) Annu Rev Biochem 54:631; b) Nilsson KGI (1988) Trends
     Biotechnol 6:256
229. a) Deonder R (1966) Methods Enzymol 8:500; b) Haynie SL, Whitesides GM (1990) Appl
     Biochem Biotechnol 23:155
230. a) Leloir LF (1971) Science 172:1299; b) Nikaido H, Hassid WZ (1971) Adv Carbohydr
     Chem Biochem 26:351

231. Kochetkov NK, Shibaev VN (1973) Adv Carb Chem Biochem 28:307
232. For the purpose of restriction L-fucose, and N-acetyl-neuramic acid are omitted in this discussion
233. Kirschning A, Floss HG unpublished results
234. a) Roseman S, Distler JJ, Moffat JG, Khorana HG (1961) J Am Chem Soc 83:659; b) Moffat JG (1966) Methods Enzymol 8:136
235. a) Nunez HA, O'Connor JV, Rosevear PR, Barker R (1981) Can J Chem 59:2086; b) Liu L-D, Liu H-W (1989) Tetrahedron Lett 30:35; c) Gokhale UB, Hindsgaul O, Palcic MM (1990) Can J Chem 68:1063
236. Kodama H, Kajihara Y, Endo T, Hashimoto H (1993) Tetrahedron Lett 34:6419
237. MacDonald DL (1962) J Org Chem 27:1107
238. Sabesan S, Neira S (1992) Carbohydr Res 223:169
239. Khwaja TA, Reese CB, Stewart JCM (1970) J Chem Soc (C) 2092
240. Ogawa T, Seta A (1982) Carbohydr Res 110:C1; b) Westerduin P, van Boom JH (1986) Tetrahedron Lett 27:1211; c) Westerduin P, Veeneman GH, van der Marel GA, van Boom JH (1986) Tetrahedron Lett 27:6271; d) Sim MM, Kondo H, Wong C-H (1993) J Am Chem Soc 115:2260; e) Klaffke W (1995) Carbohydr Res 266:285
241. a) Westerduin P, Veeneman GH, Marugg JE, van der Marel GA, van Boom JH (1986) Tetrahedron Lett 27:1211; b) Sim MM, Hirosato K, Wong C-H (1993) J Am Chem Soc 115:2260
242. Hecker SJ, Minich ML, Lackey K (1990) J Org Chem 55:4904
243. a) Schmidt RR, Stumpp M, Michel J (1982) Tetrahedron Lett 23:405; b) Schmidt RR, Stumpp M (1984) Liebigs Ann Chem 680; c) Eßwein A, Schmidt RR (1988) Liebigs Ann Chem 675; d) Hoch M, Heinz E, Schmidt RR (1989) Carbohydr Res 191:21
244. a) Kiso M, Nishihori K, Hasegawa A, Okumura H, Azuma I (1981) Carbohydr Res 95:C; b) Boons G-J, Burton A, Wyatt P (1996) Synlett 310
245. a) Bullock C, Hough L, Richardson AC (1986) Carbohydr Res 147:330; b) Adelhorst K, Whitesides GM (1993) Carbohydr Res 242:69
246. a) Withers SG, MacLennan DJ, Street IP (1986) Carbohydr Res 154:127; b) Withers SG, Percival MD, Street IP (1989) Carbohydr Res 187:43; c) Srivastava G, Alton G, Hindsgaul O (1990) Carbohydr Res 207:259; d) Leon B, Liemann S, Klaffke W (1993) J Carbohydr Chem 12:597; e) Leon B, Lindhorst TK, Rieks-Everdiking A, Klaffke W (1994) Synthesis 689
247. Niggemann J, Lindhorst TK, Walfort M, Laupichler L, Sajus H, Thiem J (1993) Carbohydr Res 246:173
248. Michalska M, Borowiecka J (1983) J Carbohydr Chem 2:99
249. a) Srivastava G, Hindsgaul O, Palcic MM (1993) Carbohydr Res 245:137; b) Inage M, Chaki H, Kusumoto S, Shiba T (1982) Chem Lett 1281
250. Wong C-H, Haynie SL, Whitesides GM (1982) J Org Chem 47:5416
251. a) Toone EJ, Whitesides GM (1991) Am Chem Soc Symp Ser 466:1; b) Heidlas JE, Lees WJ, Whitesides GM (1992) J Org Chem 57:152; c) Heidlas JE, Williams KW, Whitesides GM (1992) Acc Chem Res 25:307
252. a) Klein HW, Palm D, Helmreich EJM (1982) Biochemistry 21:6675; b) Palm D, Klein HW, Schinzel R, Buehner M, Helmreich EJM (1990) Biochemistry 29:1099
253. Evers B, Mischnick P, Thiem J (1994) Carbohydr Res 262:335
254. Zervosen A, Stein A, Adrian H, Elling L (1996) Tetrahedron 52:2395
255. Kajihara Y, Endo T, Ogasawara H, Kodama H, Hashimoto H (1995) Carbohydr Res 269:273
256. a) Nunez AH, O'Connor JV, Rosevear PR, Barker R (1981) Can J Chem 59:2086; b) Schmidt RR, Wegmann B, Jung K-H (1991) Liebigs Ann Chem 191:121; c) Ichikawa Y, Sim MM, Wong C-H (1992) J Org Chem 57:2943
257. a) Ginsburg V (1960) J Biol Chem 235:2196; b) Yamamoto K, Maruyama T, Kumagai H, Tochikura T, Seno T (1984) Agric Biol Chem 48:823; c) Ishihara H, Massaro DJ, Heath EC (1968) J Biol Chem 243:1103; d) Ishihara H, Heath EC (1968) J Biol Chem 243:1110; e) Schachter H, Ishihara H, Heath EC (1972) Methods Enzym 28:285; f) Richard WL, Serif GS (1977) Biochim Biophys Acta 484:353; g) Kilker RD, Shuey DK, Serif GS (1979) Biochim Biophys Acta 570:271; h) Butler W, Serif GS (1985) Biochim Biophys Acta 829:238

258. Lowary TL, Swiedler SJ, Hindsgaul O (1994) Carbohydr Res 256:257
259. Du M, Hindsgaul O (1996) Carbohydr Res 286:87
260. Lowary TL, Hindsgaul O (1993) Carbohydr Res 249:163
261. Haynie SL, Wong C-H, Whitesides GM (1982) J Org Chem 47:5416
262. a) Thiem J, Wiemann T (1991) Ange. Chem 103:1184; Angew Chem Int Ed Engl 30:1163;
     b) Thiem J, Wiemann (1992) Synthesis 141
263. Wong C-H, Wang R, Ichikawa Y (1992) J Org Chem 57:4343
264. Kimura T, Takayama S, Huang H, Wong C-H (1996) Angew Chem 108:2501
265. a) Martin GJ, Zhang BL, Naulet N, Martin ML (1986) J Am Chem Soc 108:5116; b) Martin
     GJ, Martin ML, Zhang B-L (1992) Plant, Cell & Environment 15:1037; c) Zhang B-L,
     Quemerais B, Martin ML, Martin GJ, Williams MJ (1994) Phytochem Analysis 5:105;
     d) Zhang B-L, Yunianta, Martin ML (1995) J Biol Chem 270:16023
266. Thiericke R, Rohr J (1993) Nat Prod Rep 10:265
267. Rouhi AM (1996) C&EN 74(39):62
268. a) Krepinsky JJ (1996) 212th ACS National Meeting, Orlando Abstract No ORGN 003;
     b) Liang R, Yan L, Loebach J, Uozumi Y, Ge M, Horan N, Gildersleeve J, Thompson C,
     Smith A, Biswas K, Sekanina K, Still WC, Kahne DE (1996) 212th ACS National Meeting,
     Orlando Abstract No ORGN 004; c) Ito Y, Kanie O, Ogawa T (1996) Angew Chem
     108:2691; Angew Chem Int Ed Engl 35:2510; d) Nicolaou KC, Winssinger N, Pastor J, De
     Roose F (1997) J Am Chem Soc 119:449; e) Douglas SP, Whitfield DM, Krepinsky JJ (1995)
     J Am Chem Soc 117:2116
269. Kanie O, Barresi F, Ding Y, Labbe J, Otter A, Forsberg LS, Ernst B, Hindsgaul O (1995)
     Angew Chem 107:2912; Angew Chem Int Ed Engl 34:2912

The Chemistry and Biology of Fatty Acid, Polyketide, and Nonribosomal Peptide Biosynthesis

Christopher W. Carreras · Rembert Pieper · Chaitan Khosla*

Department of Chemical Engineering and Chemistry, Mail Code #5025, Stanford University, Stanford, CA 94305–5025, USA
* E-mail ck@chemeng.stanford.edu

Polyketide synthases, fatty acid synthases, and non-ribosomal peptide synthetases are a structurally and mechanistically related class of enzymes that catalyze the synthesis of bio-polymers in the absence of a nucleic acid or other template. These enzymes utilize the common mechanistic feature of activating monomers for condensation via covalently-bound thioesters of phosphopantetheine prosthetic groups. The information for the sequence and length of the resulting polymer appears to be encoded entirely within the responsible proteins.

Polyketide and fatty acid biosyntheses begin with condensation of the coenzyme A thioester of a short-chain carboxylic acid "starter unit" such as acetate or propionate with the coenzyme A thioester of a dicarboxylic acid "extender unit" such as malonate or methyl malonate. The driving force for the condensation is provided by the decarboxylation of the extender unit. In the case of fatty acid synthesis, the resulting β-carbonyl is completely reduced to a methylene; however, during the synthesis of complex poly-ketides, the β-carbonyl may be left untouched or variably reduced to alcohol, olefinic, or methylene functionalities depending on the position that the extender unit will occupy in the final product. This cycle is repeated, and the number of elongation cycles is a characteristic of the enzyme catalyst. In polyketide biosynthesis, the full-length polyketide chain cyclizes in a specific manner, and is tailored by the action of additional enzymes in the pathway.

Several architectural paradigms are known for polyketide and fatty acid synthases. While the bacterial enzymes are composed of several monofunctional polypeptides which are used during each cycle of chain elongation, fatty acid and polyketide synthases in higher organisms are multifunctional proteins with an individual set of active sites dedicated to each cycle of condensation and ketoreduction. Peptide synthetases also exhibit a one-to-one correspondence between the enzyme sequence and the structure of the product. Together, these systems represent a unique mechanism for the synthesis of biopolymers in which the template and the catalyst are the same molecule.

Table of Contents

1
Introduction

Polyketides are a large family of structurally complex and pharmaceutically important molecules synthesized by the polymerization of short chain carboxylic acids such as acetate, propionate, and butyrate [1]. The term polyketide was coined over 100 years ago to refer to natural products containing multiple carbonyl and/or hydroxyl groups each separated by a methylene carbon [2]. Polyketides are typically nonessential molecules that are synthesized as secondary metabolites following the onset of stationary phase in the life cycle of an organism. Many polyketides are important therapeutic agents, including numerous antibiotics (e. g., erythromycin, tetracycline), anticancer agents (doxorubicin, enediynes), immunosuppressants (FK506, rapamycin), antiparasitic agents (avermectin, nemadectin), antifungals (amphotericin, griseofulvin), cardiovascular agents (lovastatin, compactin), and veterinary products (monensin, tylosin). In addition to bioactivity, naturally occurring polyketides can also serve alternative functions, such as being spore or flower pigments [3, 4].

Within the past few years, several excellent reviews have appeared in the literature on the genetics [5 – 7] and chemistry [1, 8 – 11] of polyketide biosynthesis. The goal of this review is to summarize our current knowledge of the biochemistry of polyketide synthases (PKSs). Particular emphasis is placed on model systems in which the *combined* application of biological and chemical techniques yields new structural and mechanistic insights into PKS function.

PKSs are enzymes that direct the biosynthesis of the carbon chain skeletons of polyketides using CoA thioesters as building blocks. Since the proposed mechanisms of PKS-catalyzed reactions are based largely on comparisons with the related fatty acid synthases (FASs), the structures and mechanisms of FASs are also reviewed here. Both polyketide and fatty acid biosynthesis initiate with the priming of the β-ketoacyl synthase (KS) with an acyl moiety "starter unit"derived from a corresponding CoA thioester (Fig. 1). Each cycle of chain elongation begins with the loading of a carboxylated "extender unit"(again, derived from a corresponding CoA thioester) onto the phosphopantetheine thiol of an acyl carrier protein (ACP). A decarboxylative condensation reaction between the two enzyme-bound acyl groups follows, resulting in the generation of an intermediate possessing a reactive β-carbonyl group. In the case of fatty acid biosynthesis, the β-carbonyl of a

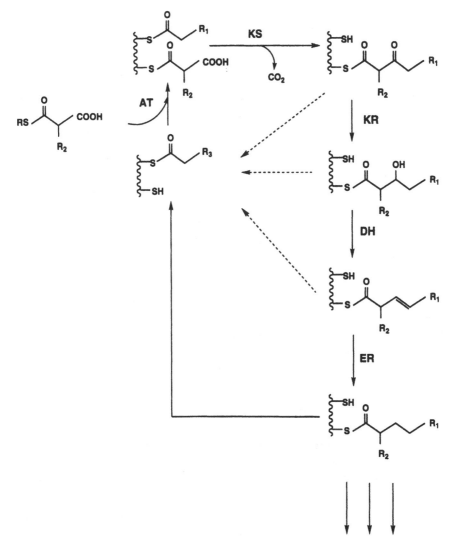

Fig. 1. The overall catalytic cycle of polyketide synthases. Within this biosynthetic scheme, different polyketide synthases can show variability with regard to the length of the polyketide chain, the choice of monomer incorporated at each step, the degree of reduction of each β-keto group, and the stereochemistry at each chiral center. For example, the *dashed arrows* illustrate how the degree of β-ketoreduction can vary at any given carbonyl

growing chain is completely reduced into a methylene; in contrast, the β-carbonyl of a growing polyketide chain may either be left unreduced or it may be stereospecifically converted into a hydroxyl, olefin, or methylene functionality. Following condensation and reduction (if it occurs), the growing chain is transferred from the ACP to the reactive cysteine of the same or different KS in preparation for another cycle of chain elongation. Thus, the final chain length, the level of ketoreduction, the stereochemistry, and the initial cyclization pattern of the full-length polyketide chain are determined by the PKS. Following formation of the initial cyclized polyketide, additional enzymes can transform the carbon skeleton with modifications such as reductions, oxidations, glycosylations, and methylations. These downstream enzymes have been reviewed elsewhere [7] and will not be considered further here.

Although precise physicochemical data concerning the structure and stoichiometry of PKSs are rather limited, cloning and DNA sequence analysis of numerous FASs and PKSs have revealed four major architectural paradigms (Fig. 2). Bacterial and plant FASs [12] and the PKSs responsible for the biosynthesis of bacterial aromatic polyketides, such as actinorhodin [13], granaticin [14], and tetracenomycin [15], are comprised of a relatively small set of (typically less than ten) active sites. Each active site is individually encoded as a distinct polypeptide. In contrast, fungal [16, 17] and animal FASs [18–21] as well as fungal PKSs, such as the 6-methylsalicylic acid synthase [22], are encoded as one or two polypeptides in which individual active sites occur as domains. In both types of multienzyme systems there are considerably fewer genetically distinguishable active sites than the total number of enzymatic reactions in the overall catalytic cycle; therefore, it has been proposed that some active sites are used iteratively in the biosynthesis of one molecule of the product. This biosynthetic strategy can be contrasted with that of the complex or "modular" PKSs, which are involved in the biosynthesis of macrolides such as erythromycin [23, 24], avermectin [25], and rapamycin [26]. Modular PKSs consist of one genetically distinguishable active site for each enzyme-catalyzed step in carbon chain assembly and modification. Active sites are clustered into modules, with each module containing a full complement of sites required for one condensation and associated reduction cycle. Finally, plant PKSs, such as the chalcone and stilbene synthases [27], are comprised of one protein which lacks any obvious sequence similarity to any of the above types of FASs or PKSs. Thus, notwithstanding similarities in the overall pathways, they appear to be an evolutionarily unrelated branch of enzymes.

Both FASs and PKSs are structurally and mechanistically related to another class of multifunctional enzymes called nonribosomal peptide synthetases. These enzymes activate amino acids as aminoacyl thioesters, which subsequently undergo condensation via formation of amide bonds, leading to biosynthesis of peptide natural products. Enzyme-bound phosphopantetheinyl groups also play a central role in the peptide assembly process. For comparison, the genetics and biochemistry of peptide synthetases are also briefly reviewed here.

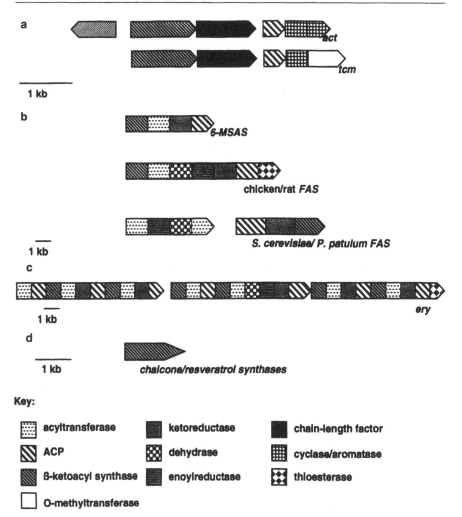

Fig. 2a–d. The major architectural paridigms of fatty acid and polyketide synthases. Relationships between genes encoding: **a** bacterial aromatic polyketide synthases, **b** eukaryotic fatty acid synthases and fungal polyketide synthases, **c** modular polyketide synthases, and **d** plant polyketide synthases

2
Fatty Acid Synthases

Fatty acids are an essential component of all living cells [28] and are used in several major cellular functions. Their structural roles include use in biological membranes, facilitating the compartmentalization that is a major feature of cellular organisms. Here, hydrophilic alcohols including choline, ethanolamine, and serine are esterified with fatty acids to create amphipathic molecules which are the primary components of biological membranes. Proteins may be acylated with fatty acids as a means of targeting these proteins to membrane-bound locations. In a quite different role, fatty acid esters of glycerol, triacylglycerols, are used as long-term storage molecules. From here, fatty acids may be oxidized to release energy, ultimately available in the form of ATP. Finally, some fatty acid derivatives, such as the prostaglandins, are used as paracrine hormones. In this case, there is an obvious parallel between fatty acid-derived hormones and biologically active polyketides.

The pathways for the synthesis and degradation of fatty acids are distinct, a typical biological strategy which allows the two pathways to be regulated independently. Similar to other metabolic pathways, NADPH is the primary redox cofactor during fatty acid biosynthesis, while NAD^+ is used during their degradation. In addition, during synthesis the growing lipid chain is processed as a thioester of a small phosphopantetheinylated protein, the ACP, while the degradative pathway uses thioesters of coenzyme A, a phosphopantetheinylated nucleotide cofactor. The enzyme activities necessary for fatty acid biosynthesis are known collectively as fatty acid synthases (FASs) [12, 29–31]. The basic reactions and associated catalytic activities in the first round of fatty acid synthesis in most organisms are shown in Table 1. Subsequent rounds of condensation and reduction involve transfer of the butyryl-ACP intermediate back to the active site thiol in the KS.

The metabolic significance of fatty acids implies that the enzyme activities involved in their synthesis are ancient and must have evolved under the selective pressure of the numerous advantages possessed by organisms with efficient fatty acid biosynthetic pathways. It is therefore no surprise that bacteria, fungi, and animals have evolved increasingly elegant solutions to the problem of coordinating these enzyme activities (Fig. 2). As summarized above, the genes encoding FASs from numerous bacteria, plants, fungi, and animals have been cloned and sequenced. Only the salient structural and mechanistic features of these enzymes will be reviewed here for comparison with PKSs. For further information and references, the reader is directed to recent reviews on the subject [12, 30–32].

Table 1. Basic reactions of fatty acid biosynthesis

Reaction	Enzyme Activity
Acetyl CoA + HCO_3^- + ATP → Malonyl CoA + ADP + P_i + H^+	Acetyl CoA carboxylase
Acetyl CoA + KS $\rightleftarrows$ Acetyl-KS + CoA-SH	Acetyl CoA: KS Transacylase
Malonyl CoA + ACP $\rightleftarrows$ Malonyl-ACP + CoA-SH	Malonyl CoA: ACP Transacylase
Acetyl-KS + Malonyl-ACP → Acetoacetyl-ACP + KS + CO_2	β-Ketoacyl-ACP Synthase (KS)
Acetoacetyl-ACP + NADPH $\rightleftarrows$ D-β-Hydroxybutyryl-ACP + $NADP^+$	β-Ketoacyl-ACP Reductase
D-β-Hydroxybutyryl-ACP $\rightleftarrows$ Crotonyl-ACP + H_2O	β-Hydroxyacyl-ACP Dehhydratase
Crotonyl-ACP + NADPH + H^+ → Butyryl-ACP + $NADP^+$	Enoyl-ACP Reductase

Higher order structural information is emerging for a variety of FASs, and it is likely that future holds models of each of these enzymes at atomic or near-atomic resolution. Electron microscopy has been useful in defining the overall shapes of the yeast [33] and chicken [34] enzymes. The fact that these enzyme complexes may be directly visualized is a striking demonstration of the massive size of these complexes. While the overall structural organization of the various components of the bacterial FAS has not yet been elucidated, atomic resolution structures of several bacterial FAS components have been determined [35–37] and others have been crystallized [38].

2.1
Acetyl CoA Carboxylase

Acetyl CoA carboxylase catalyzes the carboxylation of acetyl CoA to yield malonyl CoA, thus activating the acetate group for decarboxylative condensation onto a growing acyl chain. The activity of the bacterial enzyme is contained on four different polypeptide components, and the reaction may be divided into two half reactions, each of which requires two of these components [29]. CO_2 is coupled to biotin in the first half-reaction, then transferred from carboxybiotin to acetyl CoA in the second half-reaction. The system has been reconstituted in vitro using high concentrations of the purified components [39, 40]. In contrast to bacterial acetyl CoA carboxylase, yeast cells encode a dedicated acetyl CoA carboxylase (the *fas3* gene product) [29, 41] as a single polypeptide that forms an active tetramer which contains biotin, biotin carboxylase activity, and carboxyl transferase activity. Likewise, the animal acetyl CoA carboxylase is also a homomultimer formed from 250 kDa subunits [29, 42]. Each subunit contains domains with functions analogous to the four necessary *E. coli* proteins.

2.2
Bacterial Fatty Acid Synthases

The FAS from *E. coli* consists of numerous small proteins which must function together to carry out the reductive condensation of acetate units. Although the nature of the interactions between the full complement of protein components is not known, it is clear that various pairs of these enzymes must come into physical contact with each other during fatty acid synthesis.

The ACP is the central chaperone of fatty acid synthesis and accompanies the growing chain through each cycle of condensation [43]. The *E. coli* ACP has been cloned and overexpressed [44, 45] and its solution structure has been determined by NMR techniques [35]. The structure shows the relatively small ($M_r \approx 8.8$ kDa), acidic protein adopting an elliptical shape whose structure is dominated by several α-helices which run along the long axis of the ellipsoid. The ACP is modified at Ser-36 by a phosphodiester linkage to 4'-phosphopantetheine, the same moiety used to exchange thioesters of coenzyme A. Phosphopantetheinylation of the ACP is necessary for its function and makes thioester transfer to and from the ACP chemically similar to transfer to and from coenzyme A. This modification is accomplished by a holo-ACP synthase that obtains phosphopantetheine from coenzyme A [46].

Synthesis is initiated when an acetyl group is loaded on the β-ketoacyl-ACP synthase (KS III, the *fabH* gene product). Although it was initially postulated that this loading reaction required a specific transacylase to catalyze the acylation of KS III by acetyl CoA, it is now believed that KS III can perform this task in the absence of an additional gene product [47, 48]. Based on sequence comparisons, KS III appears unrelated to other condensing enzymes from FASs and PKSs, suggesting that the enzyme might have evolved specifically for catalyzing the priming reaction.

The synthesis of a fatty acid carbon chain is propagated using malonyl CoA, initially loaded on the phosphopantetheine moiety of the ACP, as an extender unit. Loading is accomplished through the action of the product of the *E. coli* *fabD* gene, a malonyl CoA-ACP transacylase. The crystal structure of this enzyme was recently determined [36], and although the enzyme displays a unique α/β secondary structure, the active site is reminiscent of other serine hydrolases. These enzymes share the common mechanistic feature of increasing the nucleophilicity of the active site serine oxygen through a "charge relay/ transfer" system in which a nearby histidine accepts a hydrogen bond/proton from the serine hydroxyl and donates a hydrogen bond/proton to a carbonyl oxygen (see [49, 50]). The transacylase mechanism involves the formation of a malonyl-enzyme intermediate that must resist hydrolysis by solvent water molecules. The malonyl-enzyme intermediate then docks to an ACP, which accepts the malonyl group onto its active site phosphopantetheine thiol, presumably by a similar nucleophilic displacement mechanism. Since atomic resolution structures are known for both the ACP and malonyl CoA-ACP transacylase, it may now be possible to use molecular modeling techniques to study the interaction between these FAS components.

E. coli encodes at least three (and possibly four) condensing enzymes, β-ketoacyl-ACP synthases (KSs), which catalyze decarboxylative condensations between malonyl-ACP and a growing fatty acyl chain [12, 51]. The major difference between these isoforms appears to be their chain-length specificity and differential susceptibility to covalent inhibition by the antibiotic cerulenin. As discussed above, KS III (*fabH* gene product) catalyzes the first condensation of malonate and acetate units to yield a C_4 fatty acyl-ACP. KS I (the *fabB* gene product) accepts C_2–C_{14} fatty acyl-ACP substrates, whereas KS II (the *fabF* gene product) utilizes substrates that are C_{14} or longer. Recently, the existence of a fourth condensing enzyme, KS IV (the *fabJ* product), has been proposed with a specificity between that of KS III and KS I [52]; however, another group has argued that KS IV and KS II are in fact the same proteins [53]. Some of these KSs also differentially accept various unsaturated fatty acyl-ACP substrates.

The condensation reactions occur in two steps; first the growing fatty acyl chain is transferred from the phosphopantetheine group of an ACP onto a cysteine of the KS. Subsequently, the KS binds and decarboxylates a malonyl-ACP, thereby generating a resonance-delocalized carbanion. The formation and stability of this nucleophilic species could be promoted by electrostatic or hydrogen bond interactions between the acyl carbonyl oxygen and basic groups on the enzyme [51, 54] (Fig. 3). This carbanion then attacks the carbonyl carbon of the fatty acyl-KS thioester, thus displacing the KS's cysteine sulfhydryl and forming a β-ketoacyl-ACP thioester. The β-keto group is then reduced (see below), following which a new round of elongation begins with the transfer of the growing chain back to the KS active site cysteine. KS I has been crystallized with the goal of obtaining its three-dimensional structure via X-ray crystallography [38].

The sequential action of three enzymes is required to reduce the β-keto group of a growing fatty acid. These include the NADPH-dependent β-keto-acyl-ACP reductase encoded by *fabG*, two β-hydroxyacyl-ACP dehydratases (encoded by *fabA* and *fabZ*), and an NADH-dependent enoyl-ACP reductase which is the product of the *fabI* gene [55]. These enzymes were recently purified and combined in vitro with KS III and ACP to reconstitute the first cycle of fatty acid synthesis [55]. The action of these five purified proteins was sufficient to result in the formation of butyryl-ACP from acetyl and malonyl CoA, demonstrating an understanding and ability to control the basic steps of fatty acid biosynthesis. Semiquantitative experiments using this reconstituted system demonstrated that the enoyl reductase is crucial in driving the cycle of reduction [55], since the dehydratase catalyzes an unfavorable equilibration between β-hydroxyacyl-ACP and enoyl acyl-ACP and water (see Table 1).

The crystal structure of the *Mycobacterium tuberculosis* enoyl acyl-ACP reductase bound to NADH was solved recently [37]. This enzyme is the target of the anti-tuberculosis drug isoniazid, and mutants of this enzyme are both deficient in their ability to reduce crotonyl-ACP and resistant to the activated form of this pro-drug. The core of this α/β protein is similar to the dinucleotide-binding ("Rossman") fold frequently observed in other dehydrogenases [56]. Adjacent to the nicotinamide binding site is a deep cavity lined with hydrophobic amino acid residues that is proposed to be the lipid binding site. Again, since structures of both the ACP and the enoyl reductase are known, it may now be possible to use molecular modeling techniques to identify specific sites of interaction between these FAS components.

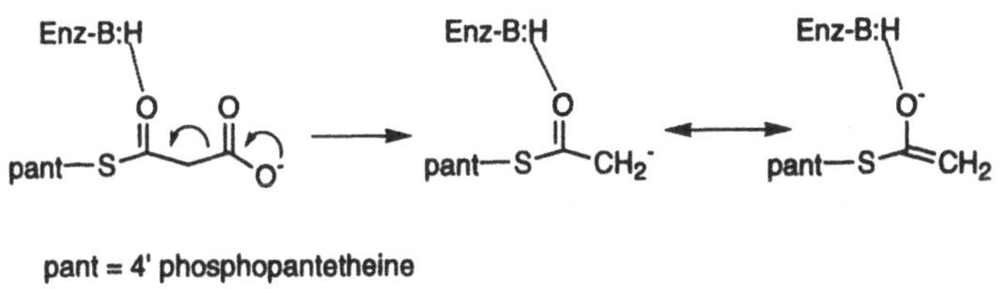

pant = 4' phosphopantetheine

Fig. 3. Decarboxylation of malonyl CoA to create the resonance-stabilized carbanion that is the nucleophilic species of the condensation reaction. A hypothetical interaction with a neutral (as shown) or positively charged basic group on the enzyme is indicated [51, 54]. Although the interaction is shown as a hydrogen bond, an equally feasible mechanism involves complete transfer of the proton

2.3
Fungal Fatty Acid Synthases

The activities involved in yeast fatty acid biosynthesis are covalently linked as separate domains of two multifunctional polypeptides, α and β, encoded by the *fas2* and *fas1* genes, respectively (Fig. 2) [57, 58]. The functionalities associated with the 220 kDa α subunit include β-ketoacyl synthase activity, β-ketoacyl reductase activity, and an ACP domain which bears a phosphopantetheinylated serine. The 208 kDa β-subunit has acetyl and malonyl CoA transacylase, palmitoyl transferase, β-hydroxyacyl-enzyme dehydratase, and enoyl acyl-enzyme reductase activities. The two subunits can be readily dissociated, and the individual activities may be measured [57].

Following the discovery that the majority of the necessary enzyme activities are contained on two polypeptides, physical methods were used to determine the stoichiometry of the complex. Early suggestions that the active complex consists of six noncovalently linked copies of each of the two necessary proteins [59] were confirmed using ultracentrifugation [60] and electron microscopic [33,60] approaches. These studies revealed an apparent molecular weight for the active, $\alpha_6 \beta_6$ complex of 2.4 million daltons, and molecular dimensions on the order of 200 Å. Moreover, electron microscopy was able to actually visualize the cylindrical shape of the complex and suggest a model for the overall organization [33]. This model suggests that the basic unit of structural organization in the yeast FAS is an $\alpha\beta$ heterodimer which forms a heterotetramer through interactions of the a subunits ($\beta\alpha\alpha\beta$). Three of these heterotetramers associate in a ring-like arrangement to form a complex that has both two- and three-fold symmetry.

The proximity of the β-ketoacyl synthase and phosphopantetheine thiols was confirmed in studies using the bifunctional reagent 1,3-dibromo-2-propanone, where the two thiols could be effectively cross-linked by the ~5 Å-long reagent [61,62]. This reagent cross-linked the two reactive thiols in such a way that two α subunits were concomitantly cross-linked. This important finding is the basis for the conclusion that the β-ketoacyl synthase/ACP active site is formed from residues derived from two different α subunits.

Once the hexameric structure of the yeast FAS was established, the number of functional active sites still remained to be determined. Earlier studies had shown that the functional complex contains approximately six equivalents each of two prosthetic groups; 4'-phosphopantetheine [60, 63], necessary for the ACP functionality, and flavin mononucleotide [64], an essential component of the enoyl reductase activity. These studies provided an early indication that each of the six active sites in the complex has a full set of the chemical groups necessary for fatty acid synthesis. Nevertheless, conflicting reports appeared in the literature as to the competence of six active sites. Whereas some reports suggested the possibility of "half-sites reactivity" (only three of the six sites are catalytically competent) [65, 66], others proposed that all six active sites could synthesize fatty acids [62]. Subsequent active site titration experiments were performed which quantitated the amount of fatty acyl products formed in the absence of turnover [67]. Single-turnover conditions were achieved through the use of

p-nitrophenyl thioacetate and thiophenyl malonate substrates, thereby depriving the FAS of CoA which is required for product release [68]. These experiments showed that each equivalent of the FAS synthesized six equivalents of fatty acyl product, indicating that the $\alpha_6 \beta_6$ FAS complex contains six (independent) chemically competent sites of fatty acid synthesis.

Three different transacylase activities are required for fatty acid synthesis in yeasts: (1) an acetyl CoA transacylase loads the priming acetyl group onto the enzyme; (2) a malonyl CoA transacylase loads malonyl extender units onto the enzyme; and (3) a palmitoyl transacylase unloads reduced acyl chains onto coenzyme A prior to their release from the enzyme. While the acetyl CoA transacylase activity is contained on a dedicated active site, malonyl and palmitoyl CoA are both transacylated via the same active site serine. Acetyl CoA is first loaded on to Ser-819 of the β subunit of the synthase, while malonyl CoA and palmitoyl CoA arrive at and depart from Ser-5421 of the β subunit [63, 69–72]. However, the S819Q(β) mutation does not abolish the acetyl CoA transacylase activity, indicating that the malonyl/palmitoyl CoA transacylase activity is also capable of processing acetate units [69]. Kinetic evidence suggests the possibility of negative cooperativity in acylation reactions which may account for early suggestions the yeast FAS is a "half-sites" enzyme [69]. Although each of the six FAS active sites is chemically competent to carry out fatty acid synthesis, the kinetics of loading the synthase with acetyl and malonyl units become less favorable as more sites are loaded. In this way, a partially loaded enzyme can reserve its remaining active sites for the processing of intermediates and products.

Following loading of acetyl and malonyl groups onto the β subunit of the enzyme, additional intramolecular transfers must occur to prepare the substrates for the decarboxylative condensation reaction which is catalyzed by the β-ketoacyl synthase domain of the α subunit. The end result of these transfers is the thioesterification of malonate by the phosphopantetheine thiol and of acetate by Cys-1305(α) of the β-keto synthase active site. This cysteine has been shown to have a dramatically lowered pK_a (<5), which would encourage its reactivity [65].

Although transfer of the acetate group from its inital site, bound as a serine ester, to its reactive position on Cys-1305(α) can occur via an intermediate in which the acetate is attached to the phosphopantetheine thiol, evidence suggests that this reaction is not kinetically competent and that the biologically significant mechanism utilizes a direct transfer from Ser-819(β) to Cys-1305(α) [73]. Transfer of the malonyl group from its initial position at Ser-5421(β) to its reactive position as a phosphopantetheine thioester occurs directly [69].

Decarboxylative condensation of the malonyl-ACP onto the β-ketosynthase-bound growing acyl chain is likely to be analogous to the corresponding reaction catalyzed by the *E. coli* β-ketoacyl synthase. Once formed, the acetoacetyl derivative remains attached to the phosphopantetheine cofactor during subsequent steps of ketoreduction, dehydration, and enoyl reduction, before the growing fatty acid is transferred to the Cys-1305 thiol in preparation for another round of elongation.

In the absence of acetyl CoA, FASs are capable of synthesizing fatty acids, suggesting that malonyl CoA can be decarboxylated and used as a starter unit (see

[74]). Elegant experiments, which capitalized on the ability of iodoacetamide to specifically alkylate the active site cysteine of the β-ketoacyl synthase, were performed, which definitively proved the capability of the yeast FAS in decarboxylating malonyl CoA [75]. Following alkylation, FAS activity is abolished; however, the enzyme still transacylates malonyl CoA to the phosphopantetheine thiol, where it is decarboxylated before being transferred back to CoA by the transacylase prior to its release as acetyl CoA.

2.4
Animal Fatty Acid Synthases

The seven activities of animal FASs are encoded as separate domains of a single 250 kDa polypeptide (Fig. 2) [30, 31]. These include a β-ketoacyl synthase, malonyl/acetyl transferase, β-ketoreductase, dehydratase, enoyl reductase, and an ACP domain with a phosphopantetheinylated serine. In addition to these activities, the animal FAS also includes a thioesterase domain which cleaves the product fatty acid from the enzyme. Proteolytic mapping of the polypeptide and genetic analysis have defined the location of the various domains in the primary sequence [30, 31].

Animal FASs are functional dimers [76]. While β-ketoacyl synthase requires dimer formation for activity [77], catalysis of the remaining FAS reactions is carried out by the monomeric enzyme. This behavior is reminiscent of yeast fatty acid synthase, where the β-ketoacyl synthase and ACP from different subunits also contribute to the same active site. Electron microscopy and small angle scattering experiments have further defined the structure of the functional complex [34, 78]. The overall shape of the molecule, as visualized by electron microscopy, is two side by side cylinders with dimensions of $160 \times 146 \times 73$ Å [34].

Similar to the yeast FAS, treatment of the chicken FAS with dibromopropanone resulted in covalent dimerization of the polypeptide, with concomitant inactivation of the enzyme [77, 79]. The enzyme groups crosslinked by dibromopropanone were shown to be cysteine and phosphopantetheine sulfhydryls from opposite subunits [79, 80]. While the β-ketoacyl synthase domain is near the NH_2-terminal of the polypeptide, the site of phosphopantetheinylation is within the ACP domain, near the enzyme's COOH-terminus. The proximity of these groups, as evidenced by their ability to be cross-linked, indicates that the two subunits of the dimer are arranged in a head-to-tail fashion which associates COOH-terminal domains from one subunit with NH_2-terminal domains from the other subunit. The notion of a head-to-tail dimer containing two sites of synthesis was further substantiated by an elegant experiment in which a hybrid dimer was formed which contained only a single competent active site [81]. Here, specific chemical modifications were performed to separately modify different reactive groups in two enzyme samples. Each of these modifications inactivated the enzyme; however, when the two chemically modified synthases were dissociated, mixed, and reassociated, FAS activity was recovered. This result demonstrates the validity of the model described above, in which the animal FAS utilizes reactive groups from each subunit at each of the two active sites.

Although the above experiments established the dimeric structure of the animal FASs, further work was necessary to establish that each of the two active sites is competent for the synthesis of fatty acids. Active site titrations, performed by inhibiting the thioesterase domain of the synthase and quantitating the bound fatty acyl products that accumulate as a result, indicated that 1 mole of fatty acyl product is formed for each mole of phosphopantetheine present [82]. Thus, each of the two subunits is chemically competent to perform all the necessary reactions of fatty acid synthesis.

Coordination of the protein domains catalyzing the individual reactions is an important aspect of catalysis by multifunctional enzymes. It has been suggested that the interdomain linker regions may have a role in the process of facilitating the movement of catalytic domains during synthesis [83]. These notions are partially based on fluorescence energy transfer experiments. While the phosphopantetheine moiety has been envisioned as a "swinging arm" that appropriately moves a growing acyl chain between the domains that catalyze the component FAS reactions, this moiety is only ~20 Å in length. Experiments were conducted in which individual active sites were labeled with fluorescent reagents and the distance between these labels was determined. The results showed that the various active sites are separated by distances substantially larger than 20 Å. For example, the distance between the ACP and thioesterase domains is estimated to be 37–48 Å [84, 85]; while the distance between the β-ketoacyl reductase and enoyl reductase sites is >49 Å [86]. In support of the head-to-tail model for organization of the overall structure, the distance between the two thioesterase domains of the dimer is >56 Å [85]. Thus, it is likely that protein dynamics play an important role in fatty acid synthesis, in the form of conformational changes that move domains relative to one another during catalysis.

A type I thioesterase domain is present at the NH_2-terminal of the animal FAS and is responsible for catalyzing hydrolysis of the completed fatty acyl chain from the enzyme. The active site contains both conserved serine and histidine residues [87] and is thought to function via a mechanism similar to that of the serine proteases [50]; however, no conserved acidic residue is present to complete the "charge relay/transfer." A second variety of thioesterase (type II) is encoded as a separate protein and interacts with the multifunctional FAS to release medium chain fatty acids [88, 89]. This enzyme has a weak sequence similarity to the type I thioesterase, which includes the conserved active site serine and histidine residues. These enzymes are also homologous to proteins encoded by genes involved in the synthesis of peptide antibiotics [90, 91] (see below).

In conclusion, the FASs are a well-studied class of enzymes which serve as models and points of reference for discussions of PKS structure and mechanisms. Each of the three classes of FAS has an analogous PKS class, and much of the detailed information concerning these PKSs is presumed through analogy to the corresponding FAS. While the iterative PKSs are analogous to bacterial FASs, the fungal and modular PKS systems resemble the multifunctional animal FASs.

3
Iterative Polyketide Synthases

3.1
Bacterial Aromatic Polyketide Synthases

Bacterial aromatic PKSs [13–15], such as the actinorhodin (*act*) PKS, are analogous to the bacterial FASs in that they are composed of several monofunctional
(and possibly bifunctional) proteins (Fig. 2). Together, these proteins control the
chain length [92,93], regiospecificity of ketoreduction [92,94,95], and regiospecificity of the initial cyclization(s) [96–99] of the nascent polyketide backbone.
A subset of three proteins, the β-ketoacyl synthase/putative acyltransferase
(KS/AT), chain length factor (CLF), and acyl carrier protein (ACP), is essential
for polyketide synthesis and comprises the "minimal" PKS [96]. The KS/AT and
CLF are homologous to one another (although the CLF lacks the active site
cysteine and serine, typically present in the KS and putative AT domains,
respectively), and it is possible that these two proteins associate as heterodimers
or heteromultimers. In addition, gene clusters encoding bacterial aromatic PKSs
include open reading frames (ORFs) encoding ketoreductases (KRs), cyclases
(CYCs), and aromatases (AROs), which act on the nascent polyketide backbone
to guide its folding into specific structures.

The genes encoding aromatic PKSs were initially isolated on the basis of their
ability to complement mutant strains that were blocked in their ability to produce polyketides [100]. DNA sequencing of the *tcm* [15], *gra* [14] *and act* [13]
PKS gene clusters revealed ORFs with a high level of sequence similarity to
the ACP and KS components of bacterial FASs. Likewise, the KR was initially
recognized by virtue of its similarity with a bacterial ribitol dehydrogenase
[101]. In contrast, the function of other PKS genes could not be deduced from
sequence comparison studies.

An experimental approach was developed to analyze the function of these
and other PKS components. It involved heterologous expression of defined PKS
gene sets in an engineered strain of *Streptomyces coelicolor* from which the
entire *act* gene cluster was deleted. Spectroscopic and isotope labeling analysis
of the reporter polyketide products yielded insights into properties of the
recombinant PKSs [92]. These studies utilized homologous gene sets from
several aromatic PKS clusters from different species, each of which directs
biosynthesis of its own natural product intermediate. Analysis of products made
by truncated PKSs as well as hybrid PKSs provided insights into the function
and molecular recognition features of individual subunits. Specifically, the
following catalytic properties (see, for example, Fig. 4) have been investigated
using this approach:

1. *Chain length*: Polyketide carbon chain length is dictated by the minimal PKS
   [92, 93, 96]. Within the minimal PKS, the ACP can be interchanged without
   affecting specificity, whereas the CLF is crucial. Although some KS/CLF combinations are functional, others are not, therefore, biosynthesis of a polyketide chain of specified length can be ensured with a minimal PKS in which

both the KS and CLF originate from the same PKS gene cluster. So far, chain lengths of 16 (octaketide), 18 (nonaketide), 20 (decaketide), and 24 carbons (dodecaketide) have been generated with minimal PKSs from the *act, fren, tcm,* and *whiE* PKS clusters, respectively. At least in the presence of a KR, some minimal PKSs, such as the *fren* and the *whie* minimal PKSs, show relaxed chain length control [93, 102].

2. *Ketoreduction*: Ketoreduction requires a KR. The *act* KR (the only one studied so far) can catalyze reduction of the C-9 carbonyl (counting from the carboxyl end) of any length nascent polyketide backbone studied so far [92, 93, 102]. Furthermore, the *act* KR is compatable with all the minimal PKSs mentioned above. Homologous KRs have been identified in other PKS clusters. These enzymes may also catalyze ketoreduction at C-9 since all the corresponding natural products undergo this modification. In unusual circumstances, C-7 ketoreductions have also been observed with the *act* KR [93].

3. *Cyclization of the first ring*: Although the minimal PKS alone can provide partial control for the formation of the first ring, the regiospecific course of this reaction may be influenced by other PKS proteins. For example, when present alone, most minimal PKSs studied thus far produce a reporter polyketide with a C-7/C-12 cyclization; however, alternatively cyclized products are also observed. For example, the *tcm* minimal PKS alone generates both C-7/C-12 and C-9/C-14 cyclized products [96]. The presence of certain accessory subunits can result in considerable restriction of this variability in cyclization regiochemistry. For example, a KR with any minimal PKS restricts the nascent polyketide chain to cyclize exclusively with respect to the position of ketoreduction: C-7/C-12 cyclization for C-9 ketoreduction and C-5/C-10 cyclization for C-7 ketoreduction [92, 93]. Likewise, use of the TcmN enzyme alters the regiospecificity to C-9/C-14 cyclizations for unreduced polyketides of different lengths, but has no effect on reduced molecules [98, 99]. Finally, although the *act* ARO normally recognizes a reduced polyketide as a substrate, its presence with the minimal *act* PKS results in substantial enrichment of the C-7/C-12 cyclized product over an aberrantly cyclized shunt product [103].

4. *First ring aromatization*: The first ring in unreduced polyketides aromatizes noncatalytically. In contrast, an aromatase is required for reduced polyketides [97]. There appears to be a hierarchy in the chain length specificity of these subunits from different PKS clusters. For example, the *act* ARO will recognize only 16-carbon chains, the *fren* ARO recognizes both 16- and 18-carbon chains, while the *gris* ARO recognizes chains of 16, 18, and 20 carbons [104].

5. *Second ring cyclization*: C-5/C-14 cyclization of the second ring of reduced polyketides may be achieved with an appropriate cyclase [97, 104]. While the *act* CYC can cyclize octa- and nonaketides, it does not recognize longer chains. No equivalent C-5/C-14 CYC with specificity for decaketides or longer chains has been identified, although the structures of natural products such as jadomycin and griseusin imply their existence. In the case of sufficiently long unreduced chains with a C-9/C-14 first ring cyclization, formation of a C-7/C-16 second ring is at least partially influenced by the minimal PKS [96], and is further controlled by tcmN [98, 99, 105].

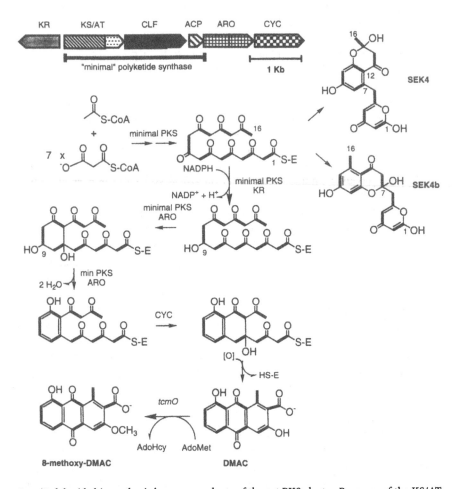

Fig. 4. Polyketide biosynthesis by gene products of the act PKS cluster. Presence of the KS/AT, CLF, and ACP is sufficient for the production of two 16-carbon polyketides, SEK4 and SEK4b both in vivo [103] and in vitro [107]. In the presence of the *act* ketoreductase (KR), aromatase (ARO) and cyclase (CYC), the octaketide intermediate is converted into DMAC. DMAC can be converted into 8-methoxy DMAC both in vivo and in vitro through the S-adenosylmethionine (Adomet)-dependent action of the tcmO methyltransferase [207]

An understanding of the mechanisms by which aromatic PKSs synthesize a highly labile poly-β-ketone intermediate of precise chain length and guide it towards a regiospecifically reduced and cyclized product presents a formidable biochemical challenge. As summarized above, mutagenesis and heterologous expression of recombinant bacterial aromatic PKSs have provided some insight into the functions and molecular recognition features of the different protein components of these enzymes. However, our knowledge regarding the fundamental mechanisms of aromatic PKS function remain rudimentary at best. In large measure this has been due to the absence of fully active cell-free systems for the study of these enzymes. Recently however, the situation has changed [106, 107], and these in vitro systems are expected to facilitate the purification of the responsible proteins, followed by their kinetic, physicochemical, and ultimately structural characterization.

The work of Shen and Hutchinson on the tetracenomycin (*tcm*) PKS led to the development of the first cell-free system for aromatic polyketide biosynthesis [106]. These authors demonstrated the conversion of acetyl and malonyl CoA into Tcm F2, the first 20-carbon intermediate in the biosynthesis of tetracenomycin, in crude extracts prepared from a *Streptomyces glaucescens* host bearing plasmids expressing genes for the minimal *tcm* PKS, and either the tcmN or tcmJ cyclases. Synthesis of Tcm F2 was authenticated by its conversion to Tcm F1 and Tcm D3 by the highly specific action of purified preparations of the Tcm F2 cyclase and Tcm F1 monooxygenase [108–110]. The results support a model in which the three proteins of the minimal PKS are responsible for synthesis of a nascent, enzyme-bound linear decaketide that is subsequently folded and cyclized by the *tcm*N and possibly *tcm*J gene products.

In addition to demonstrating the first cell-free PKS activity, early studies of the *tcm* PKS provided information concerning the properties of isolated aromatic PKSs. The first suggestion that a PKS is capable of decarboxylating malonyl CoA to acetyl CoA for use as a starter unit came from studies which used extracts containing the *tcm* PKS. These extracts catalyzed polyketide synthesis in the absence of acetyl CoA with [^{14}C]malonyl CoA as the sole substrate. Although it remains possible that an additional enzyme present in the extract is responsible for this activity, decarboxylation of malonyl CoA for use as a starter unit has previously been demonstrated in purified FAS systems, as indicated above, and this result is the first indication that PKSs may also have this capability.

More recently, similar experiments were reported using a system in which genes for the actinorhodin (*act*) PKS were expressed in a recombinant strain of *S. coelicolor* from which the entire *act* gene cluster had been deleted [107]. This work demonstrated the efficient conversion of acetyl and malonyl CoA into two aromatic polyketide products, SEK 4 and SEK4b, by cell-free preparations of the *act* minimal PKS (Fig. 4). Micromole quantities of polyketides were synthesized in vitro, indicating that the method may be useful for the synthesis of products that are not directly obtainable using fermentation methods. The system showed several similarities to the *tcm*PKS system described above: synthesis did not require acetyl CoA and occurred using [^{14}C]malonyl CoA as the sole substrate, suggesting that the *act* minimal PKS may also be capable of decarboxylating extender units for use as starter units. In addition, synthesis could be inhibited by the addition of cerulenin or reagents which specifically modify sulfhydryl groups.

The activity of the complete *act* PKS has also been demonstrated. Extracts containing the *act* KR, CYC and ARO in addition to those of the minimal PKS catalyzed the synthesis of 3,8-dihydroxy-1-methylanthraquinone-2-carboxylic acid (DMAC), the primary in vivo product of the complete *act* PKS. The structure of this product was verified by its enzymatic conversion into 8-methoxy DMAC by the tcmO O-methyltransferase, which was prepared from a separate strain which expressed this protein (Fig. 4).

The purification of bacterial aromatic PKSs is likely to result in elucidation of at least some of the structural features of these enzymes. For example, knowledge of the stoichiometry of the protein components in the active complex could suggest possible physical mechanisms for the chain elongation steps and perhaps imply a mechanism for the role of the CLF in control of the chain lengths of polyketides synthesized by these enzymes. Towards this end, several ACP proteins have been overexpressed and purified to homogeneity [111, 112]. Likewise, purification of KR, ARO, and CYC components and reconstitution of their activities with the *act* minimal PKS will be useful in answering questions regarding their precise functions and the temporal sequence of the reduction and cyclization reactions, which may or may not occur before completion of the synthesis of the polyketide backbone.

3.2
Fungal Polyketide Synthases

Cell-free systems capable of in vitro synthesis of 6-methylsalicylic acid (6-MSA) and a related tetraketide, orsellinic acid, were developed long before the advent of recombinant DNA technologies in the field of natural product biosynthesis [113–115] (Fig. 5). Since then, the biosynthetic mechanisms and molecular recognition features of 6-methylsalicylic acid synthase (6-MSAS) have been extensively studied. 6-MSAS initiates synthesis with an acetyl group derived from acetyl CoA, extends the polyketide chain to a tetraketide via three decarboxylative condensations of malonyl CoA-derived extender units, and uses NADPH to specifically reduce one of resulting carbonyls to a hydroxyl group. In its natural producer, *Penicillium patulum*, the product, 6-MSA is subsequently glycosylated to form the antibiotic patulin [116].

More recently, a PKS involved in the biosynthesis of $(4R)$-$4[(E)$-2-butenyl]-4-methyl-L-threonine (Bmt), one of the constituent amino acids of the methylated undecapeptide cyclosporin A, has also been the subject of biochemical studies using cell-free systems. Cyclosporin is produced by the fungus *Tolypocladium niveum*. Bmt is known to be one of the crucial residues for the interaction of cyclosporin A with its natural ligand cyclophilin [117]. ^{13}C-labeled acetate was shown to be incorporated in vivo into Bmt in an ^{13}C-enrichment pattern suggesting a polyketide pathway. Additionally, in vivo feeding with ^{13}C-labeled methionine resulted in an enriched ^{13}C-peak attributed to the 4-methyl side chain in Bmt. This indicated the biosynthetic origin of the methyl group from S-adenosyl-L-methionine (AdoMet) [118]. The first phase of the biosynthesis of Bmt thus involves the assembly of a tetraketide intermediate including three cycles of reductive and dehydrative steps at the emerging β-keto groups. This was confirmed by the isolation of $3(R)$-hydroxy-$4(R)$-methyl-$6(E)$-octenoic acid via incorporation of $[1$-$^{13}C, ^{18}O_2]$acetate in vivo. In a cell-free assay, partially purified protein fractions from *T. niveum* were tested for polyketide synthase activity including the substrates malonyl CoA, $[1$-$^{14}C]$acetyl-CoA, NADPH, and S-$[^{14}C$-methyl]AdoMet. $3(R)$-hydroxy-$4(R)$-methyl-$6(E)$-octenoic and its C2-C3 dehydrated analog were isolated by HPLC and identified by LC-MS [119]. The enzymatic conversion of

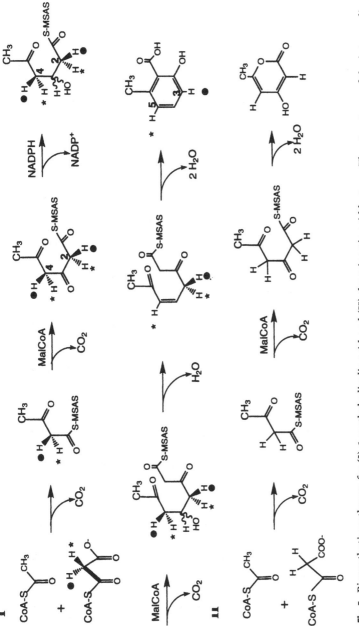

Fig. 5. Biosynthetic pathways for (I) 6-methylsalicylic acid and (II) the triacetic acid lactone. The structures of the intermediates have not been identified. The stereochemical course of the prochiral carbons (C-2 and C-4 in the triketide intermediate, C-3 and C-5 in 6-MSA) was investigated using (*R*)- and (*S*)- [1-¹³C, 2-²H]malonic acid extender substrate analogs in a coupled assay with 6-MSAS and succinyl-CoA transferase. The distinguishable hydrogens originating from the chiral malonyl CoA are labeled with *H and H. Triacetic acid lactone synthesis is catalyzed by 6-MSAS in the absence NADPH

3(R)-hydroxy-4(R)-methyl-6(E)-octenoic acid or its thioester into the 2-amino derivative remains to be analyzed.

The 6-MSAS gene (Fig. 2) was isolated using an immunological screening of a genomic *P. patulum* DNA expression library [22]. Sequencing of the 5322 base pair (bp) ORF showed that it encodes a 190 kDa protein, in agreement with the molecular weight of purified 6-MSAS determined by SDS-PAGE. Comparison with the bacterial and animal FAS as well as aromatic PKS sequences revealed several homologous regions, tentatively identified as acyltransferase, β-ketoacyl synthase, acyl carrier protein, and ketoreductase domains. The central and COOH-terminal portions of the 6-MSAS gene did not show homology to any FAS or PKS components and are currently of unknown function. The order of the functional domains is similar to the order of domains in animal FASs and modular PKSs, but different from the arrangement found in the yeast FAS. 6-MSAS, bacterial FASs, animal FASs, and aromatic PKSs use their protein components in an "iterative" fashion, in which a single set of domains is capable of performing all of the necessary chain elongation steps. Unlike animal FASs, 6-MSAS does not harbor a thioesterase domain.

Gel filtration of purified 6-MSAS indicates that it is a 750 kDa homotetramer [120]. When treated with 1,3-dibromopropanone (DBP), 6-MSAS behaves similarly to animal FASs and inactivation occurs concomitantly with its cross-linking to a covalent homodimer. By analogy to the well-characterized DBP cross-linked yeast and animal FASs [61,62,79,80], DBP is suggested to cross-link active site sulfhydryl residues of the β-ketoacyl synthase cysteine and the ACP pantetheine. This notion is supported by the observation that preincubation with either acetyl or malonyl CoA precludes cross-linking. Thus, the functional significance of the tetrameric assembly of 6-MSAS remains a mystery.

6-MSAS from *P. patulum* was separated from the FAS via sucrose gradient centrifugation [121,122] and thus shown to constitute a distinct multifunctional enzymatic system. It was purified to homogeneity and found to be a 190 kDa multifunctional enzyme [22,120]. The enzyme was more stable in the presence of its substrates and at mildly basic pH values. The pH optimum of the enzyme was 7.6 and apparent K_m values for its substrates were 10 µM (acetyl-CoA), 7 µM (malonyl CoA), and 12 µM (NADPH) [115,120,123]. The rate for triacetate lactone formation in the absence of NADPH was determined to be ten-fold lower than for 6-MSA formation (Fig. 5) [120]. Analogous to FASs and peptide synthetases, 4'-phosphopantetheine is a covalently bound cofactor of 6-MSAS [124]. Likewise, iodoacetamide and *N*-ethylmaleimide were found to inactivate the enzyme, suggesting the presence of catalytic sulfhydryl residues in 6-MSAS [124]. Furthermore, in the presence of malonyl CoA and NADPH, low concentrations of iodoacetamide convert 6-MSAS into a malonyl CoA decarboxylase. Without external addition of acetyl-CoA, 6-MSAS decarboxylates the malonyl group and the derived acetyl moiety is used as a starter unit for the formation of 6-MSA [125].

Phenylmethylsulfonyl fluoride, a serine protease inhibitor which also inhibits the thioesterase function of animal FASs, does not inactivate 6-MSAS, implying a lack of a thioesterase function in 6-MSAS. 6-MSA release from the enzyme does not appear to follow a thioesterase (serine)-catalyzed mechanism [120], nor is 6-MSA released from the enzyme as a CoA derivative [32]. Thus, a poly-ketide release mechanism similar to bacterial FASs and yeast FAS, which contain a bifunctional transacylase for covalent activation of the malonyl group and for acyl transfer of the palmitoyl thioester product to CoA, may be excluded. It is therefore unclear how 6-MSAS catalyzes the cleavage of the thioester bond to its final intermediate. (A similar question arises for other iterative PKSs.) In con-trast to 6-MSAS, however, the release of the Bmt-polyketide precursor occurs as a CoA thioester, which is reminiscent of FASs of fungal origin (see above).

Propionyl CoA could be incorporated as a primer unit analog into 6-ethyl-salicylic acid at an eight-fold reduced rate, indicating relaxed starter unit speci-ficty [125]. Incorporation of additional starter unit analogs by 6-MSAS was also recently reported [126]. GC/MS analysis indicated that while the levels of buty-ryl-, crotonyl-, hexanoyl-, and heptanoyl incorporation from the CoA derivatives into the respective 6-MSA analogs were less than 5% of acetyl incorporation, the yields of triketide lactones derived from the long-chain primer unit analogs, catalyzed by 6-MSAS in the absence of NADPH, were 14%–74%. A high speci-ficity of the β-ketoacyl reductase, but a relaxed specificity of the acyltransferase and β-ketoacyl synthase for chain length intermediates was deduced from these results. A triacetic acid ethyl ester, used as an intermediate analog, was shown to be a substrate for the 6-MSAS-ketoreductase [124], suggesting that ketore-duction occurs on the triketide intermediate. Triacetic acid ethyl ester is an effi-cient substrate for the ketoreductase site of 6-MSAS (K_m of 140 µM). In an expe-riment to elucidate the order of reactions following the ketoreduction of the six-carbon intermediate, 5-oxo-2,3-hexenyl ethyl ester was chemically synthesized as a reference substance. If the elimination of the hydroxyl group occured imme-diately after ketoreduction, the ^{3}H-labeled hexenyl ethyl ester might be formed incubating 6-MSAS with ^{3}H-labeled NADPH and the triacetic acid ethyl ester. No hexenyl ethyl ester was detected, and it was inferred that the third conden-sation yielding a tetraketide intermediate with a C-5 hydroxyl group preceeds the C4/C5 dehydration reaction. It remains to be established, however, whether the ethyl ester analog of this intermediate is recognized by 6-MSAS.

6-MSA does not contain any chiral carbon centers. Before the aromatization of the six-membered ring occurs, two prochiral carbons (C-2 and C-4 in the six-carbon intermediate) evolve, each of which loses a hydrogen in the process of the dehydratization/aromatization steps. In addition, C-3 of the six-carbon inter-mediate forms a chiral center when the ketone is reduced to a hydroxyl by a keto-reductase activity (Fig. 5). The chirality of this hydroxyl carbon is unclear since the intermediate has not been isolated. It is also unknown if this carbon retains its chirality in an eight-carbon intermediate or whether the hydroxyl is elimi-nated by dehydration prior to the third condensation reaction. The stereospeci-ficity at the prochiral C-2 and C-4 carbons in the reaction intermediates was addressed using chemically synthesized (R)- and (S)-[1-^{13}C, 2-^{2}H]malonate precursors which were enzymatically converted into CoA derivatives via succinyl CoA transferase [127, 128]. Thus, the prochiral methylene in malonyl CoA was "replaced" by chiral, double-labeled (S)- or (R)-[1-^{13}C, 2-^{2}H]malonyl CoA substrates in the reaction mixture with 6-MSAS. The condensation is expected to occur with inversion of configuration and the intact methylene hydrogens are incorporated into the polyketide chain. The pattern of "heavy atoms" retained and mass spectroscopic analysis of the 6-MSA products support a mechanism in which the two hydrogen atoms (at C-2 and C-4 in the six-carbon intermediate, at C-3, C-5 in 6-MSA) with opposite absolute orientations are specifically removed during the processing of the intermediates by dehydration reactions. A coupled assay including purified 6-MSAS and succinyl CoA transferase was found to minimize exchange (racemization) of the methylene hydrogens of the chiral malonyl CoAs. (R)- and (S)-$[1$-^{13}C, 2-^{2}H]malonate, converted to malonyl CoA by succinyl CoA transferase, were turned over in situ by 6-MSAS. By replacing acetyl CoA with acetoacetyl CoA in the assay mixture and thus omitting one condensation cycle, only one of the two hydrogens (C-3 in 6-MSA) is derived from isotope-labeled, chiral malonyl CoA and turned over by 6-MSAS. Mass spectroscopic results suggested that the hydrogen retained at this position is derived from H_{Re} in malonyl CoA, respectively H_{Si} in the intermediate (Fig. 5).

Similar stereochemical studies have also been conducted on the orsellinic acid synthase from *Penicillium cyclopium*, a multisubunit enzyme composed of a 130 kDa protein [129, 130]. The catalytic cycle of this PKS is identical to the 6-MSAS cycle, except that it lacks any ketoreduction or dehydration reactions. Unlike 6-MSAS, enolizations occurring during orsellinic acid biosynthesis are not stereospecific.

Using stereospecifically tritiated NADPH enantiomers, it was demonstrated that only tritium from the $[4$-$^{3}H_{Si}]$NADPH is incorporated into 6-MSA [32]. Presumably, the ketoreductase possesses stereospecificity for the H_{Si} of the dihydronicotinamide ring of NADPH and chain length specificity for the six-carbon ketide intermediate.

Recently, the 6-MSAS gene has been reconstructed to facilitate optimal expression in *S. coelicolor*. Transformation of *S. coelicolor* with an expression plasmid carrying the 6-MSAS gene yielded strains that produced 6-MSA [131]. Manipulations of the 6-MSAS gene should lead to the analysis of the functions attributed to certain regions of the multifunctional enzyme on behalf of sequence similarities. Site-directed mutagenesis of highly conserved residues might reveal mechanistic details during the catalytic processing of 6-MSA polyketide intermediates.

3.3
Plant Polyketide Synthases

Chalcone and stilbene synthases are related plant PKSs [132]. Chalcones, such as naringenin chalcone, are produced as the biosynthetic precursors of flavinoids, while stilbenes are produced for their antifungal properties. Plant PKSs are likely to have evolved independently from any of the aforementioned PKS and FAS systems [133, 134] and are atypical in many respects. These homodimeric enzymes consist of a single ~40 kDa gene product (Fig. 2) [135]. The two active sites of the dimer function independently of one another [136]. Plant PKSs lack an ACP component, are not phosphopantetheinlyated, and act directly on CoA thioesters [134, 137].

The peanut chalcone synthase and parsley stilbene synthases have been cloned, expressed in *E. coli*, and purified to homogeneity [135, 137]. The enzymes appear to be mechanistically similar; each catalyzes the formation of a tetraketide from three molecules of malonyl CoA that are decarboxylated and condensed with a starter unit derived from *p*-coumaroyl CoA or a similar CoA thioester (Fig. 6). No reductions or dehydrations occur during either chalcone or stilbene synthesis, and some products spontaneously cyclize following their release from the enzyme. A major feature that distinguishes chalcone and stilbene synthases is that the latter perform an additional decarboxylation to remove a carbon atom that is present in chalcone products [132, 138]. The presence of this additional carboxyl group results in a different cyclization pattern for chalcone products. The precise mechanisms by which chalcone and stilbene synthases determine the fate of this carbon atom are not known.

Substrates are processed directly through cysteine thiols, and cerulenin and iodoacetamide are able to inhibit the activity of plant PKSs by modifying sulfhydryl groups of these residues [134]. Although chalcone and stilbene synthases do not share a high level of overall sequence similarity with FASs and PKSs, the cysteine that is modified by cerulenin is conserved, suggesting that the active sites may be structurally related [51].

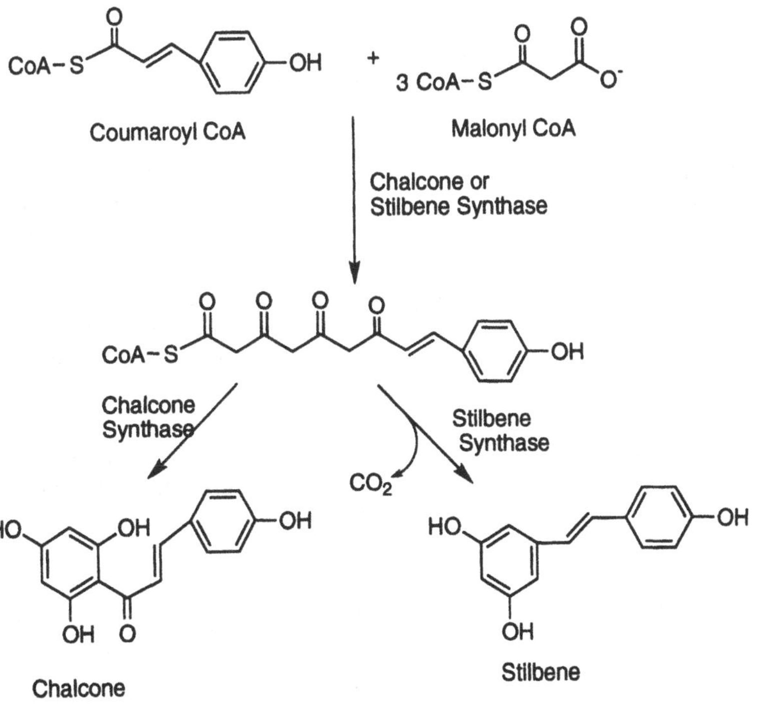

Fig. 6. Reactions catalyzed by chalcone and stilbene synthases. Each enzyme condenses three malonyl CoA extender units onto *p*-coumaroyl-CoA. Stilbene synthases catalyze an additional decarboxylation, resulting in a different pattern of cyclization for chalcone versus stilbene products

Most biologically relevant reactions of chalcone and stilbene synthases use p-coumaroyl CoA as a starter unit for synthesis; however, CoA thioesters of acetate, butyrate, hexanoate, benzoate, cinnamoate and phenylpropionate are also accepted [137, 138]. Although the molecular details of the interaction of the synthase with starter units have not been elucidated, the identification of a single amino acid substitution that changes the starter unit specificity of a stilbene synthase suggests genetic engineering as a promising route to novel chalcone and stilbene derivatives [27].

The small size, availability of purified enzymes from several sources, and apparent simplicity of plant PKS systems make them ideal targets for detailed structural studies, including X-ray crystallogaphy. However, since these enzymes represent a unique solution to the problem of condensing CoA thioesters, it remains to be seen whether an improved understanding of the mechanisms of plant PKSs will aid the engineering of other PKS systems.

4
Modular Polyketide Synthases

4.1
6-Deoxyerythronolide B Synthase

Erythromycin, a clinically important antibiotic, is a prominent example of a glycosylated macrocyclic polyhydroxyoxolactone (macrolide). Its aglycone, 6-deoxyerythronolide B (6-dEB), reveals a regular structure of seven recurring three-carbon units, as originally postulated by Woodward and Gerzon [139, 140]. Evidence for the origin of 6-dEB from propionate precursors was obtained by in vivo isotope-labeling studies in the erythromycin producer *Saccharopolyspora erythraea* [141, 142]. Based on the polyketide paradigm, 6-dEB was suggested to be synthesized via repetitive decarboxylative condensations between a propionyl-CoA primer and six methylmalonyl CoA extenders. Using a variety of ^{13}C-, ^{18}O-, ^{2}H-, and multiple-labeled substrates and advanced intermediate analogs, Cane and coworkers concluded that the biosynthesis of 6-dEB involves a processive mechanism in which the final reduction state and the stereochemistry of each of the carbons is set directly after the condensation reaction between the growing intermediate and a methylmalonyl extender [143–147]. Notwithstanding these advances, direct insights into the structure and properties of the enzymes responsible for 6-dEB biosynthesis did not emerge until the cloning and DNA sequencing of the erythromycin gene cluster [24].

6-dEB biosynthesis involves three contiguous ORFs of approximately 10 kb each, encoding three large multidomain proteins, designated deoxyerythronolide B synthase (DEBS) 1, 2, and 3 [23, 24]. Detailed sequence comparisons revealed that each of these proteins consists of eight to ten domains with considerable sequence similarity to enzymes responsible for each of the individual steps of fatty acid biosynthesis. Moreover, these domains are arranged such that each protein contains two functional units or modules, each of which carries all the

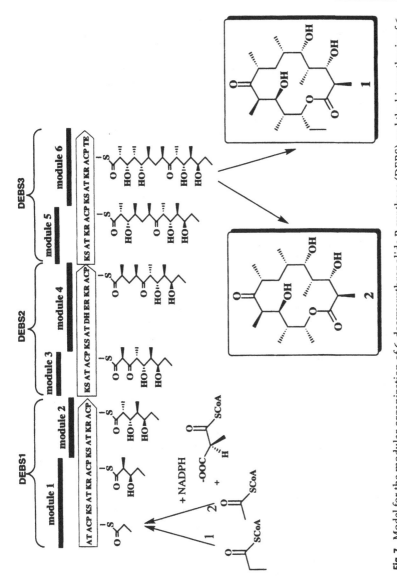

Fig. 7. Model for the modular organization of 6-deoxyerythronolide B synthase (DEBS) and the biosynthesis of 6-deoxyerythronolide B (1) and 8,8a-deoxyoleandolide (2) by DEBS [208]. Each of the six modules accounts for the one polyketide chain extension and β-ketoreduction cycle. The active sites are designated as follows: acyltransferase (*AT*), β-ketoacyl-ACP transferase (*KS*), acyl carrier protein (*ACP*), β-ketoreductase (*KR*), dehydratase (*DH*), enoylreductase (*ER*), and thioesterase (*TE*)

requisite catalytic activities for one of six cycles of polyketide chain elongation and reductive modification of the resultant β-ketoacyl thioester. The availability of the structural genes for 6-DEBS has provided the conceptual framework for the design of many of the most important experiments carried out on this truly remarkable multienzyme system.

According to the now widely accepted model of Katz and coworkers (Fig. 7), the "loading" acyltransferase (AT-L) domain at the NH$_2$-terminal of DEBS 1 initiates the polyketide chain-building process by transferring the propionyl-CoA primer unit via the pantetheinyl residue of the first acyl carrier protein

(ACP-L) domain to the active site cysteine of the ketosynthase of module 1 (KS1). The acyltransferase in module 1 (AT1) loads methylmalonyl CoA onto the phosphopantetheine thiol of the ACP domain of module 1. KS1 then catalyzes the first polyketide chain elongation reaction by decarboxylative acylation of the methylmalonyl residue by the propionyl starter unit, resulting in the formation of a 2-methyl-3-ketopentanoyl-ACP thioester. This intermediate is then reduced by the ketoreductase of module 1 (KR1), giving rise to enzyme-bound (2S,3R)-2-methyl-3-hydroxypentanoyl-ACP. At this point, module 1 has finished its task and the diketide product is transferred to the active site cysteine of KS2, where it undergoes another round of condensation and reduction, resulting in the formation of the corresponding triketide. This process is repeated several times, with each module being responsible for a separate round of polyketide chain elongation and reduction, as appropriate, of the resulting β-ketoacyl thioester. Finally, the thioesterase (TE) at the COOH-terminal of DEBS 3 is thought to catalyze release of the finished polyketide chain by lactonization of the product generated by module 6. The sequential order of active sites (domains) is in complete agreement with a model in which the growing polyketide chain moves along the enzyme template and is reminiscent of the thiotemplate model for nonribosomal peptide synthesis (see below).

The DEBS proteins were initially purified from S. erythraea [148]. Their molecular weights agreed well with the predicted molecular weights of the DEBS proteins: DEBS 1, 370 kDa; DEBS 2, 380 kDa; DEBS 3, 330 kDa. NH_2-terminal sequence analysis and polyclonal antibodies, raised against recombinant protein fragments of COOH-terminal regions of DEBS 2 and DEBS 3, were used to prove their identity. A truncated gene encoding COOH-terminal DEBS 3 didomain, including the ACP6 and the thioesterase, was expressed in E. coli, although the ACP was found to lack a phosphopantetheinyl group. This was also found to be the case with the complete recombinant DEBS proteins obtained from an E. coli expression system; however, the E. coli-derived proteins could be acylated with propionyl-CoA and methylmalonyl CoA [149]. Remarkably, although both R and S configurations are observed among the six methyl-branched centers of 6-dEB, only the (2S)-enantiomer of methylmalonyl CoA was found to be an extender substrate for all condensation reactions during the pathway [150]. Thus, it was concluded that variations in stereospecificity at the methyl-branched carbons must be introduced either immediately before β-ketocondensation at the methylmalonyl-thioester stage or after condensation, at the ketide-thioester stage. The active site and mechanism for the epimerization catalysis remains to be identified.

Limited proteolysis with various proteases of the three DEBS proteins revealed that proteolytic cleavage occurs preferentially at domain boundaries. In the case of DEBS 1, for example, cleavage was observed between the loading domain (AT-L/ACP-L), the first two domains of module 1(KS1/AT1), the following two domains of module 1 (KR1/ACP1), and the entire module 2 [151]. Similar patterns of intermodular cleavage were also seen in the cases of DEBS 2 and DEBS 3 [152]. Purified modules of DEBS 1, DEBS 2, and DEBS 3 obtained by limited proteolysis were found to behave as homodimers in gel filtration and sedimentation equilibrium experiments. Furthermore, chemical cross-linking using the bifunctional sulfhydryl cross-linker 1,3-dibromopropanone resulted in the isolation of a cross-linked homodimer of module 5 [152]. In contrast, a protein fragment containing module 6 but lacking its ACP domain could not be cross-linked, suggesting that the ACP and the KS from a given module are likely to be juxtaposed in a head-to-tail fashion. Based on these results, a model has been proposed for the structure of the multifunctional three-protein DEBS complex in which the KSs, ACPs, and ATs form a rod-like core in a double helical assembly of modules [152].

Following the cloning and sequencing of the DEBS gene cluster, Katz and coworkers performed two important knock-out mutagenesis experiments. In one, the ketoreductase domain of module 5 was inactivated, and the predicted 5-oxo-6-deoxyerythronolide B was isolated from the mutant strain [24]. From this, it could be inferred that the active sites within module 6 can recognize and process the 12-carbon oxo-intermediate. Likewise, mutagenesis of a conserved motif in the enoylreductase domain of module 4 resulted in production of the expected C6-C7 anhydro-macrolactone [153]. These pioneering experiments demonstrated the intrinsic tolerance of active sites in DEBS with respect to molecular recognition of advanced intermediate analogs.

In a different approach, the DEBS gene cluster was expressed in a *S. coelicolor* based host-vector system [154]. The complete DEBS not only produced 6-dEB but also 8,8a-deoxyoleandolide, an analogous 14-membered macrolactone featuring a methyl side chain at C13 which is biosynthetically derived from an acetate starter unit (Fig. 7), indicating that the enzyme had a relatively relaxed specificity towards starter units. The three DEBS proteins could be isolated in significantly higher quantities from the recombinant strain than from wild-type *S. erythraea*. In contrast to the expression of DEBS enzymes in *E. coli*, post-translational modification of the ACP domains with the pantetheine cofactors was catalyzed in the heterologous host by a holo-ACP synthase activity. In an attempt to probe the individual function of DEBS 1 alone, the DEBS 2 and DEBS 3 genes were deleted from this heterologous expression system. The resulting mutant produced $(2R,3S,4S,5R)$-2,4-dimethyl-3,5-dihydroxy-n-hexanoic acid δ lactone (C_9-lactone), the propionate-derived product expected from a lactonization reaction of a triketide intermediate [155] (Fig. 8). In order to investigate the thioesterase function with respect to its catalytic properties and recognition of polyketide intermediates, two groups successfully repositioned the thioesterase from the end of module 6 to the end of module 2, thus constructing a fusion protein between DEBS 1 and the thioesterase designated DEBS 1+TE [156, 157]. The recombined multienzymes, one engineered in *S. erythraea*, the other in *S. coelicolor*, were shown to synthesize the C_9-lactone. The increased yield of this product over the amount produced by DEBS1 alone suggests that the thioesterase catalyzes lactonization of the truncated polyketide intermediate.

A new 12-membered macrolactone, $(8S,9S)$-8,9-dihydro-8-methyl-9-hydroxy-10-deoxymethynolide was isolated by fusing the ACP6-thioesterase didomain to the COOH-terminal end of KR5 (Fig. 8). In addition to demonstrating that the TE-catalyzed cyclization reaction involves regiospecific recognition of the same terminal hydroxyl on chains of different lengths, the experiment also illustrates the feasibility of constructing hybrid modules. In contrast, in the presence of the

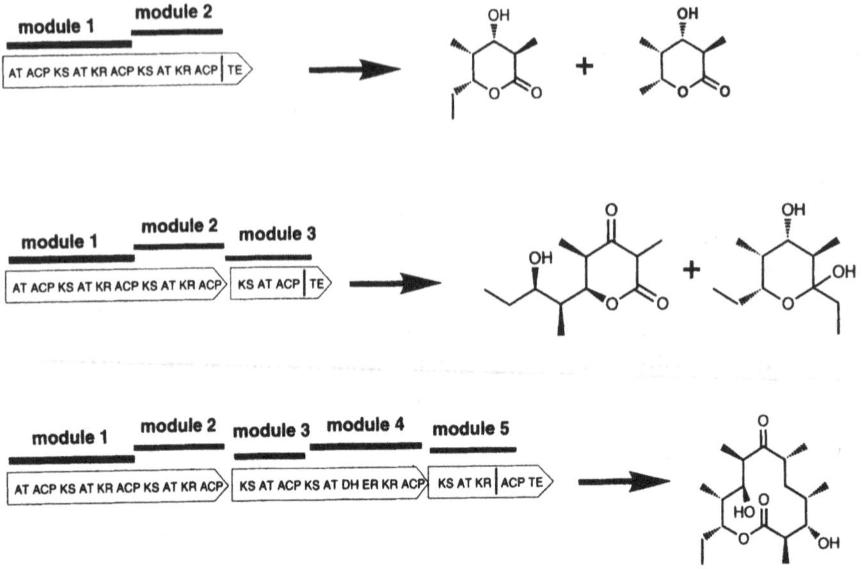

Fig. 8. Polyketides produced by *S. coelicolor* strains expressing truncated forms of DEBS [157, 158]. Structures were determined by isotope enrichment and NMR spectroscopy

TE domain, the tetraketide intermediate from the third DEBS module gave rise to CK13a and CK13b as primary and secondary biosynthetic products [158]. CK13a is a 6-membered lactone whereas CK13b is a decarboxylated hemiketal. These results illustrate how intermediates of the 6-deoxyerythronolide B pathway that do not undergo DEBS-catalyzed macrolactonization can cyclize into structurally diverse products. Further manipulation of DEBS along these lines could help elucidate the mechanisms by which DEBS determines the stereochemistry of the chiral carbon centers, how polyketide chain transfer reactions are mediated, and the significance of linker regions between the conserved domains.

The addition of in vitro methods to the above in vivo strategies for studying the mechanisms of modular PKSs has been a long sought-after goal. However, until recently, the challenge of synthesizing complex polyketides in vitro has represented a major technical barrier in the field of polyketide biosynthesis. With the development of improved expression systems for DEBS and the availability of genetic tools to truncate the DEBS gene cluster, cell-free studies on fully active DEBS multienzymes have recently become feasible. DEBS-catalyzed in vitro synthesis of 6-dEB and 8,8a-deoxyoleandolide was recently demonstrated using [1-^{14}C]propionyl-CoA, methylmalonyl CoA, and NADPH as substrates [159]. In vitro synthesis of the C$_9$-lactone, which required only DEBS 1+TE, has also been demonstrated [159, 160]. In fact, it was more efficient as judged from the unambiguous identification of the product via ^{13}C-NMR analysis, using [1-^{13}C]propionyl-CoA as the labeled substrate [159]. Synthesis was completely inhibited by thiol inhibitors such as iodoacetamide, *N*-ethylmaleimide, and cerulenin, which are also known to irreversibly inactivate FASs.

DEBS 1+TE was also shown to efficiently incorporate an advanced chain elongation intermediate, $(2S,3R)$-2-methyl-3-hydroxypentanoyl-N-acetylcysteamine (NAC) thioester, into the triketide lactone [159]. $(2S,3R)$-2-methyl-hydroxybutyryl-NAC is also turned over by DEBS 1+TE and converted into the C_8-lactone, normally derived from an acetate starter and methylmalonate extenders [161] (Fig. 9). Thus, it is evident that these NAC-thioesters are treated as diketide intermediates, not as starter substrates by the enzyme, suggesting that the substitution pattern and not the chain length is crucial for the recognition of substrates by DEBS.

DEBS appears to have a fairly broad starter unit specificity. Butyryl-CoA and acetyl-CoA are incorporated into the respective C_{10}- and C_8-lactones, although at significantly lower yields than propionyl-CoA [160, 161] (Fig. 9). Remarkably, DEBS 1 + TE can process unreduced and partially reduced intermediates as well. The AT-L in the NH_2-terminal part of DEBS 1 covalently activates the starter units with comparable efficiency [161]. In contrast, malonyl CoA is not recognized as an extender unit by acyltransferase domains (R. Pieper, unpublished). When NADPH is excluded from the reaction mixture, a pyran-2-one was synthesized by DEBS 1 + TE [161]. Consistent with this result, incubation of $(2S,3R)$-2-methyl-hydroxypentanoyl-NAC thioester and methylmalonyl CoA with DEBS

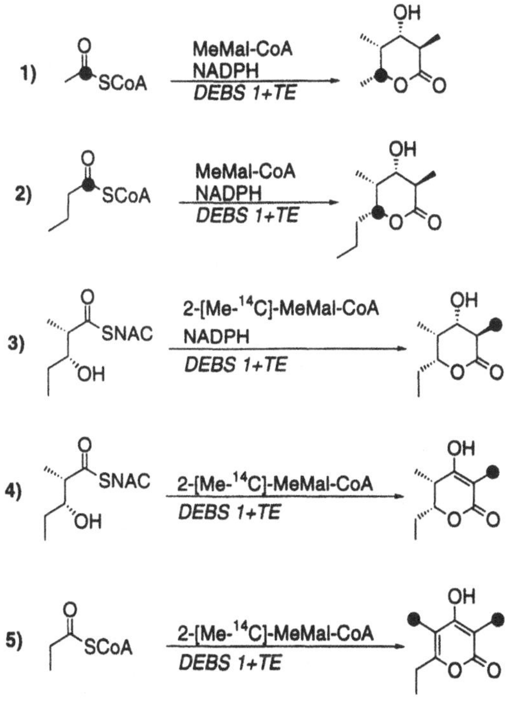

Fig. 9. Polyketide products generated via in vitro biosynthesis by the fusion protein DEBS 1+TE. *1* and *2*, ^{14}C-labeled alternate starter unit substrates were used to synthesize the C_8- and C_{10}-lactones, respectively. *3* and *4*, $(2S,3R)$-2-methyl-3-hydroxypentanoyl-N-acetylcysteamine (NAC) thioester was used as an intermediate to synthesize the C_9-lactone and the 3-oxo analog, respectively. *5*, In the absence of NADPH, a pyran-2-one is generated using propionyl-CoA and ^{14}C-labeled methylmalonyl CoA

1+TE in the absence of NADPH resulted in synthesis of the 3-oxo analog of the C_9-lactone [162]. These experiments reconfirmed the tolerance of condensation sites of DEBS for less reduced intermediates.

The overall k_{cat} for the synthesis of the C_9-lactone by DEBS 1+TE was measured to be 3.4 min^{-1} [163]. For 6-dEB synthesis by complete DEBS, an apparent k_{cat} of 0.5 min^{-1} was determined. The measured k_{cat} of DEBS 1+TE indicates that this truncated PKS is highly active in a cell-free system and approaches catalytic activity comparable to in vivo levels. The apparent K_m for (2S) methylmalonyl CoA consumption by DEBS 1+TE is 24 µM. Although starter units with shorter and longer side chains are incorporated into the respective triketide δ lactones, DEBS 1+TE has a 7.5-fold preference for propionyl-CoA over butyryl-CoA and a 32-fold preference over acetyl-CoA. In the absence of the primer propionyl-CoA, DEBS 1+TE turns over (2S)-methylmalonyl CoA and NADPH at the same rate as in the presence of the starter unit. This suggests that DEBS 1+TE decarboxylates the extender unit, transfers the evolving propionyl group to its active site KS1, and thus initiates polyketide chain elongation.

4.2
Other Modular Polyketide Synthases

Gene clusters (or parts thereof) controlling the biosynthesis of several other complex polyketides, including avermectin [25], rapamycin [26], oleandomycin [164], and soraphen [165], have been isolated and sequenced. In all cases, the PKSs have been found to be organized into individual modules with each module containing the appropriate sets of active sites. Thus, the modular hypothesis appears to be well-substantiated now in several model systems. Intriguingly, in the case of rapamycin, biosynthesis of the entire macrocycle involves activity of a 12-module PKS as well as a peptide synthetase module (see below).

5
Nonribosomal Peptide Synthetases

Nonribosomal peptide synthesis is based on the catalytic activity of multifunctional enzyme systems that resemble FASs and PKSs. Like polyketides, non-ribosomally synthesized peptides are a structurally diverse family with an impressive range of biological activities. Peptide synthetases have been isolated from a variety of bacteria, filamentous fungi, and plants. Their products are synthesized in the late exponential phase under growth-limiting conditions. Observed structural variations include: (1) peptide chain lengths of 1–20 amino acid components, (2) unusual precursors such as hydroxy acids, chromophores, D-amino acids, and L-amino acids not found in ribosomal proteins, (3) peptide modifications (N-methylation, acylation, glycosylation, epimerization reactions), and (4) overall structure (linear, cyclic, lactonized, branched-chain, depsi-, peptides) (Fig. 10). Biological activities range from antimicrobial agents (gramicidin S, β-lactams), biosurfactants (surfactin), siderophores (enterobactin), to animal or plant toxins (HC-toxin) (for detailed overviews see [166–169]). Interestingly,

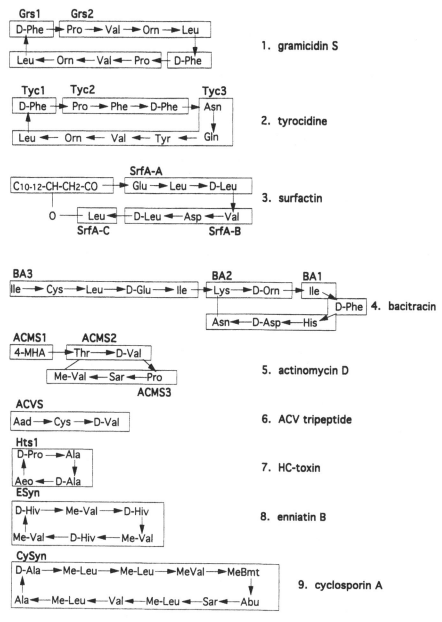

Fig. 10. Structures of nonribosomally synthesized peptides of bacterial origin (*1–6*) and fungal origin (*7–9*). *Me*, N-methylated peptide bond; *Orn*, ornithine; *4-MHA*, 4-methyl-3-hydroxyanthranilic acid; *Aad*, aminoadipic acid; *Aeo*, 2-amino-9,10-epoxy-8-oxodecanoic acid; D-*Hiv*, D-hydroxyisovaleric acid; *Bmt*, (4R)-4-[(E)-2-butenyl]-4-methyl-L-threonine; *Abu*, α-aminoisobutyric acid; *Sar*, sarcosine. The *boxes* signify gene products for peptide synthetases composed of modules which activate and process the indicated amino acids

structurally diverse compounds like the undecapeptide cyclosporin A and the complex polyketide FK506 bind similar biological target molecules, the immunophilins, and interfere with identical T cell signaling pathways.

The catalytic mechanisms and molecular recognition properties of peptide synthetases have been studied for several decades [169]. Nonribosomal peptides are assembled on a polyenzyme-protein template, first postulated by Lipmann [170]. The polyenzyme model was refined into the thiotemplate mechanism (Fig. 11) in which the amino acid substrates are covalently bound via thioester linkages to active site sulfhydryls of the enzyme and condensed via a processive mechanism involving a 4′-phosphopantetheine carrier [171–173]. The presence of a covalently attached pantetheine cofactor was first established in a cell-free system that catalyzed enzymatic synthesis of the decapeptides gramicidin S and tyrocidine. As in the case of fatty acid synthesis, its role in binding and translocating the intermediate peptides was analyzed [174, 175].

The ACV synthetase gene was the first complete peptide synthetase gene to be isolated and sequenced [176, 177]. Sequence comparisons revealed that the ACV

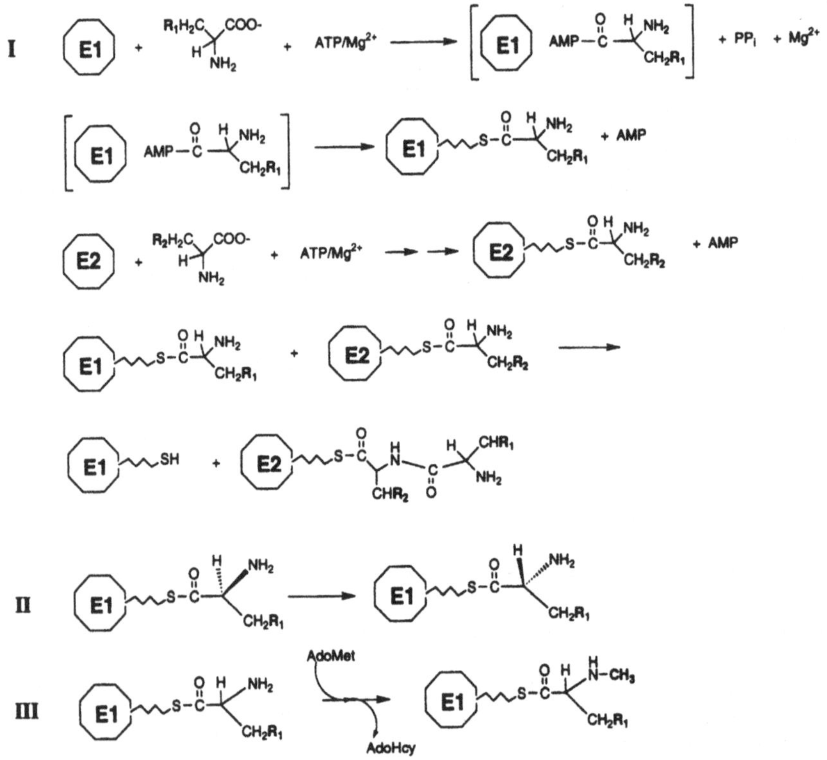

Fig. 11. Reaction scheme for (I) substrate amino acid activations and dipeptide formation, (II) racemization, and (III) N-methylation. *E1* and *E2* are symbols for enzyme activities (from peptide synthetase modules) either on the same or separate proteins. *R1* and *R2* are amino acid side chains, where *R1* is part of the first amino acid activated by the peptide synthetase indicates the linkage between 4′-phosphopantetheinylated enzyme and the substrate or peptide intermediate. *AdoMet*, S-adenosylmethionine; *AdoHcy*, S-adenosylhomocysteine

synthetase is composed of three regions sharing a high degree of sequence similarity with each other as well as with sequences of related enzymes, such as gramicidin S synthetase 1 (Grs1) [90], tyrocidine synthetase 1 (Tyc1) [178], and several types of adenylate-forming enzymes. Furthermore, the number of regions in the ACV synthetase corresponds to the number of constituent amino acids in the product, δ-(L-aminoadipyl)-L-cysteinyl-D-valine. This was the first evidence for modular arrangement of peptide synthetases. Since then, the genes encoding several peptide synthetases have been cloned and sequenced (for review, see [169, 179]). In all cases this modular organization is preserved. Perhaps the most striking example of a multienzyme system is the cyclosporin synthetase, with a molecular weight of 1690 kDa, arranged in 11 modules and incorporating 11 amino acids into its product. As illustrated in Fig. 12, sequence comparison of individual modules also established the existence of conserved domains; the functions of these domains have been further analyzed by mutagenesis and heterologous expression. For example, Grs1, a single module polypeptide (1098 amino acids) involved in gramicidin S biosynthesis, consists of a 556 amino acid NH_2-terminal domain that is sufficient for phenylalanyl adenylation. An additional 100 amino acids downstream of this domain are required for L-phenylalanyl thioester formation, but not for racemization of the amino acid. Deletion of the COOH-terminal 291 amino acids in the module aborted phenylalanyl epimerization, suggesting that epimerization activity lies in this portion of a module [180]. Finally, N-methyltransferase activity, when present, is found in the central part of the module [181, 182].

Each module in a peptide synthetase contains a conserved region of about 600 amino acid residues, which is involved in adenylation and subsequent thioester formation of an amino acid (Fig. 12). The amino acid sequence around the pantetheinylation site also shows significant similarity to mammalian FASs and modular PKSs. Site-directed mutagenesis of the conserved serine in this region of Tyc1 (Fig. 12) resulted in loss of the covalent activation of its substrate L-Phe. This finding supports the idea that the serine is the crucial amino acid-binding site [183]. Thus, as in the case of mammalian FASs and modular PKSs, the domains in peptide synthetases harboring the pantetheine cofactor are parts of larger proteins. (An exception to this rule occurs in the case of D-alanine incorporation into D-alanyl-lipoteichoic acid, in which a distinct ACP protein covalently binds the substrate [184]; this is reminiscent of the dissociated aromatic PKSs. However, unlike the conserved active site cysteine in the β-ketoacyl synthase domain of FASs and PKSs, peptide synthetases lack a second universally conserved thiol in their modules.

An interesting variation on the theme of one pantetheine per extender unit might occur in the case of the enniatin synthetase, in which a third pantetheine binding motif has been identified via sequence comparison [185]. Since enniatin biosynthesis involves repeated condensation of a dipeptide intermediate, it has been postulated that the third pantetheine binding site represents a "waiting position" for peptidol intermediates, so that additional dipeptidol units can be formed using the two other pantetheine groups.

Prior to their transfer to active site sulfhydryls, individual amino acids are activated by the corresponding modules as noncovalently bound adenylates.

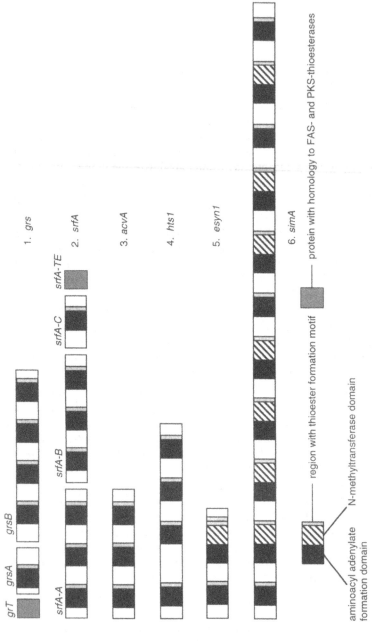

Fig. 12. Organization of peptide synthetases in a distinct number and order of modules. *1–3*, Peptide synthetases encoded by the bacterial operons *grs*, *srfA*, and *acvA*; *4–6*, peptide synthetases encoded by the fungal operons *hts1*, *esyn1*, and *simA*. *grT* and *srfA-TE* signify genes putatively coding for thioesterases. Further details are included in the text. Each of the modules can be dissected into two or three conserved domains responsible for adenylate formation (*black*), *N*-methyltransfer (*striped*), substrate thioester binding (*gray*). Regions with yet unknown functions (*white*)

Activation of carboxylic acids as adenylates at the expense of ATP also occurs in various other enzyme classes, including the aminoacyl-tRNA synthetases and acyl-CoA ligases (e.g., 4-coumaryl-CoA ligase, long chain fatty acid synthase, firefly luciferase). Whereas there is no apparent structural similarity to tRNA synthetases, sequence comparisons with acyl-CoA ligases have facilitated the definition of adenylate-forming domains spanning 500 amino acid residues in peptide synthetase modules [186] (Fig. 12). Consistent with this notion, point mutations in the corresponding domains of gramicidin synthetase 2 (Grs2), surfactin synthetase (SrfA-B), and Tyc1 have been identified which resulted in a reduction or abolition of adenylation activity [183, 187–189]. Likewise, tryptic Tyc1 peptide fragments, photoaffinity-labeled with the ATP analog 2-azidoadenosine triphosphate, mapped to sequences within this domain [190]. In contrast to the adenylation domains of individual modules, the domains involved in the peptide bond formation or in peptidyl transfer have not been identified.

The thiotemplate mechanism (Fig. 11) was originally proposed to account for three crucial observations: (1) individual amino acid substrates are activated as thioesters, (2) the number of active sites is consistent with the number of constituent amino acids in the peptide, and (3) a single phosphopantetheine moiety was postulated to act as a swinging arm, which connects the active sites and extends the peptide chain. Based on this model, a multidomain structure was postulated for peptide synthetases with substrate activation sites on separate domains. This mechanism, which was initially proposed for gramicidin S synthesis [166], was revised as a multiple phosphopantetheine carrier model [191, 192]. In the revised model the constituent amino acids are activated as aminoacyl adenylates, with the hydrolysis of pyrophosphate from ATP being the driving force for the reaction. A sulfhydryl from phosphopantetheine is then acylated with the amino acid. Modifications such as the racemization or AdoMet-dependent N-methylation of an amino acid occur subsequently at a thioester-bound stage. Once the first and second amino acids are activated, dipeptide formation occurs via aminolysis of the first aminoacyl thioester by the free amine of the second amino acid. This cycle continues until the full length-peptide is assembled. The release mechanism for the peptide is not yet understood. Cyclization or hydrolysis might occur spontaneously or be catalyzed by a thioesterase functionality.

Like modular PKSs, peptide synthetases also epimerize some substrates and/or intermediates. For example, the starter substrate amino acid of cyclosporin A is D-Ala. Racemization of alanine is not catalyzed by an integrated subunit of cyclosporin A synthetase, but by alanine racemase. This is a separate, pyridoxal phosphate-dependent enzyme [193]. In contrast, Grs1 and Tyc1 covalently activate L-Phe as a thioester and subsequently epimerize the amino acid [194]. D-Phe is the only epimer accepted as a substrate for dipeptide formation by Grs2 and Tyc2 [195, 196]. No racemization activity is detected in a pantetheine-deficient mutant of Grs1 [197]. Deletion mutagenesis pointed to the requirement of the COOH-terminal part of the module for epimerizing L-Phe to D-Phe [180]. In contrast, the biosynthesis of actinomycin D, a bicyclic chromopentapeptide lactone (Fig. 10), involves formation of the dipeptide 6-MHA (methylanthranilic acid)-L-Thr-L-Val prior to epimerization of the L-Val extender to D-Val [198]. Using chemically synthesized p-toluyl-dipeptide analogs it was shown that dipeptide diastereomers including either L-Val or D-Val could be isolated as activated intermediates from the actinomycin synthetase.

Some modules of eukaryotic peptide synthetases, such as the cyclosporin synthetase and the enniatin synthetase, catalyze the formation of N-methylated peptides using S-adenosyl-L-methionine (AdoMet) as the methyl donor. N-methylation of amino acids occurs at the thioester-bound stage [199] and is catalyzed by integrated N-methyltransferase domains within the relevant modules [181, 182, 185]. As shown in Fig. 12, the COOH-terminal region of the valine-activating module of enniatin synthetase includes a 430 amino acid region that does not show homology to bacterial peptide synthetase modules but to methyltransferases. Recombinant enniatin synthetase fragments expressing this domain in E. coli were shown to be photoaffinity-labeled with AdoMet. A proteolytic 45 kDa enniatin synthetase fragment photoaffinity-labeled with AdoMet exactly matched the region assigned to the methyltransferase domain [200].

In contrast to mammalian FASs and modular PKSs, thioesterase functions are not integrated parts of peptide synthetases. However, a 29 kDa protein encoded by a gene at the 5'-end of the grs operon (coding for gramicidin S synthetase) revealed strong homology with thioesterase regions of PKSs [90]. Similar genes were found to be associated with the srfA operon (coding for the surfactin synthetase) and a gene cluster responsible for the biosynthesis of the tripeptide bialaphos [91, 201]. Although the genes have been shown to be essential for in vivo production of the antibiotic peptides, the thioesterase proteins themselves do not have to be included in a cell-free peptide synthesis system. Rather, the synthetases alone terminate peptide synthesis effectively by cyclization or hydrolysis.

Although most modules of peptide synthetases show a strict specificity towards natural amino acids, some modules can incorporate alternative amino acids into their products. For example, the enniatin synthetase shows a relatively broad substrate specificity for hydrophobic amino acids (L-Val, L-Leu, L-Ile). Both nonmethylated and methylated versions can be processed to generate depsipeptides in a cell-free system [202]. Likewise, cyclosporins with a variety of alternative amino acids have also been synthesized from cell-free cyclosporin synthetase preparations[203]. Various enniatins and cyclosporins can also be isolated from fermentation broths [204, 205].

Finally, as in the case of PKSs, the feasibility of generating hybrid peptide synthetases via genetic engineering has been demonstrated recently. A strategy was developed to substitute gene portions coding for individual modules of peptide synthetases into a distinct segment of the engineered peptide synthetase operon srfA [206]. The srfA gene expresses surfactin synthetase in Bacillus subtilis. Surfactin is an acyl-peptidolactone composed of a β-hydroxy fatty acid side chain and seven amino acid residues, including a leucine at position 7 in srfA-C (Fig. 10). Modules of ACV and Grs2 synthetases were thus shown to replace the leucine-activating module of srfA-C, resulting in engineered biosynthesis of novel surfactins.

6
Summary and Perspectives

In summary, PKSs, FASs, and peptide synthetases are remarkable examples of related multienzyme systems that catalyze the formation of structurally diverse and complex organic molecules via building-block mechanisms. In each case a relatively small repertoire of reactions is used iteratively, and the structural features of each building-block as well as the growing intermediate are controlled with elegant precision. Although the overall process is reminiscent of the biosynthesis of nucleic acids and ribosomal proteins, Nature seems to have evolved a rather different strategy in these cases. Instead of modularizing a template molecule, the multifunctional enzymes themselves appear to be modularized.

In contrast to template-based systems such as the replication, transcription, and translation machineries, our understanding of the structure and mechanisms of PKSs, FASs, and peptide synthetases is rudimentary at best. However, as reviewed here, the past decade has witnessed a dramatic confluence of chemical and biological methods for studying these problems. This has resulted in not only a substantially improved understanding of these nontemplate enzymes, but in an increased ability to productively engineer these systems for the biosynthesis of novel products. While it may be early to predict the ultimate impact of these amazing biocatalysts in molecular science and engineering, given the rich history of natural products in biology, chemistry, and medicine, one can look forward to equally exciting opportunities with "unnatural" natural products.

Acknowledgements. Research in the authors' laboratory was supported by the National Science Foundation (MCB9417419 and an NSF Young Investigator Award to CK), the National Institutes of Health (CA66736), a David and Lucile Packard Fellowship (CK), and by gifts from Merck and Co. and Sankyo Co. CWC was a recipient of postdoctoral fellowships from the Ford Foundation and the National Institutes of Health (1 F32 GM17543-01).

7
References

1. O'Hagan D (1995) Nat Prod Rep 12:1
2. Collie N (1893) J Chem Soc 122
3. Davis NK, Chater KF (1990) Mol Microbiol 4:1679
4. Durbin ML, Learn GH, Huttley GA, Clegg MT (1995) Proc Natl Acad Sci, USA 92:3338
5. Hopwood DA, Khosla C (1992) Ciba Found Symp 171:88
6. Katz L, Donadio S (1993) Annu Rev Microbiol 47:875
7. Hutchinson CR, Fujii I (1995) Annu Rev Microbiol 49:201
8. O'Hagan D (1991)The Polyketide Metabolites. Ellis Horwood, Chichester, UK
9. O'Hagan D (1992) Nat Prod Rep 9:447
10. O'Hagan D (1993) Nat Prod Rep 10:593
11. Simpson TJ (1995) Chemistry & Industry 11:407
12. Magnuson K, Jackowski S, Rock CO, Cronan JEJ (1993) Microbiol Rev 57:522
13. Fernández-Moreno MA, Martínez E, Boto L, Hopwood DA, Malpartida F (1992) J Biol Chem 267:19278
14. Sherman DH, Malpartida F, Bibb MJ, Kieser HM, Bibb MJ, Hopwood DA (1989) EMBO J 8:2717
15. Bibb MJ, Biró S, Motamedi H, Collins JF, Hutchinson CR (1989) EMBO J 8:2727

16. Chirala SS, Kuziora MA, Spector DM, Wakil SJ (1987) J Biol Chem 262:4231
17. Mohamed AH, Chirala SS, Mody NH, Huang WY, Wakil SJ (1988) J Biol Chem 263:12315
18. Holzer KP, Liu W, Hammes GG (1990) Proc Natl Acad Sci, USA 86:4523
19. Witkowski A, Rangan VS, Randhawa ZI, Amy CM, Smith S (1991) Eur J Biochem 198:571
20. Schweizer M, Takabayashi K, Laux T, Beck K-F, Schreglmann R (1989) Nuc Acid Res 17: 567
21. Amy C, Witkowski A, Naggert J, Williams B, Randhawa Z, Smith S (1989) Proc Natl Acad Sci, USA 86:3114
22. Beck J, Ripka S, Siegner A, Schiltz E, Schweizer E (1990) Eur J Biochem 192:487
23. Cortes J, Haydock SF, Roberts GA, Bevitt DJ, Leadlay PF (1990) Nature 348:176
24. Donadio S, Staver MJ, J.B. M, Swanson SJ, Katz L (1991) Science 252:675
25. MacNeil DJ, Occi JL, Gewain KM, MacNeil T, Gibbons PH, Ruby CL, Danis SJ (1992) Gene 115:119
26. Schwecke T, Aparacio JF, Molnar I, König A, Khaw L, Haydock SF, Oliynyk M, Caffrey P, Cortes J, Lester JB, Böhm GA, Staunton J, Leadlay PF (1995) Proc Natl Acad Sci, USA 92:7839
27. Schröder G, Schröder J (1992) J Biol Chem 267:20558
28. Stryer L (1995) Biochemistry. 4th edn, WH Freeman, New York
29. Wakil SJ, Stoops JK, Joshi VC (1983) Ann Rev Biochem 52:537
30. Smith S (1994) FASEB J 8:1248
31. Wakil S (1989) Biochemistry 28:4523
32. Schorr R, Mittag M, Müller G, Schweizer E (1994) J Plant Phisiol 143:407
33. Stoops JK, Kolodziej SJ, Schroeter JP, Bretaudiere JP (1992) Proc Natl Acad Sci, USA 89:6585
34. Stoops JK, Wakil SJ, Uberbacher EC, Bunick GJ (1987) J Biol Chem 262:10246
35. Holak TA, Nilges M, Prestegard JH, Grononborn AM, Clore GM (1988) Eur J Biochem 179:9
36. Serre L, Verbree EC, Dauter Z, Stuitje AR, Derewenda ZS (1995) J Biol Chem 270:12961
37. Dessen A, Quémard A, Blanchard JS, Jacobs WRJ, Sacchettini JC (1995) Science 267:1638
38. Olsen JG, Kadziola A, Siggaard-Andersen M, Chuck J, Larsen S, von Wettstein-Knowles P (1995) Prot Pep Lett 1:246
39. Guchhait RB, Polakis E, Dimroth P, Stoll E, Moss J, Lane MD (1974) J Biol Chem 249:6633
40. Polakis SE, Guchhait RB, Zwergel EE, Lane MD, Cooper TG (1974) J Biol Chem 244:6657
41. Al-Feel W, Chirala SS, Wakil SJ (1992) Proc Natl Acad Sci, USA 89:4534
42. Takai T, Yokoyama C, Wada K, Tanabe T (1988) J Biol Chem 263:2651
43. Vagelos PR (1973) Acyl group transfer (acyl carrier protein). In: The enzymes. Academic, New York, p 155
44. Rawlings M, Cronan JEJ (1992) J Biol Chem 267:5751
45. Jones AL, Kille P, Dancer JE, Harwood JL (1993) Biochem Soc Trans 21:202S
46. Lambalot RH, Walsh CT (1995) J Biol Chem 270:24658
47. Jackowski S, Murphy CM, Cronan JEJ, Rock CO (1989) J Biol Chem 264:7624
48. Jackowski S, Rock CO (1987) J Biol Chem 262:7927
49. Stroud RM (1974) Sci Amer 231:24
50. Perona JJ, Craik CS (1995) Prot Sci 4:337
51. Siggaard-Andersen M (1993) Prot Seq Data Anal 5:325
52. Siggaard-Andersen M, Wissenbach M, Chuck J, Svendsen I, Olsen JG, von Wettstein-Knowles P (1994) Proc Natl Acad Sci, USA 91:11027
53. Magnuson K, Carey MR, Cronan JE (1995) J Bact 177:3593
54. Dewar MJS, Dieter KM (1988) Biochem 27:3302
55. Heath RJ, Rock CO (1995) J Biol Chem 270:26538
56. Rossman MG, Liljas A, Granden C, Banaszak L (1975) In: The enzymes. Academic, New York, p 61
57. Stoops JK, Wakil SJ (1978) Biochem Biophys Res Comm 84:225
58. Schweizer E, Werkmeister K, Jain J (1978) Mol Cell Biochem 21:95
59. Kresze GB, Oesterhelt D, Lynen F, Castorph H, Schweizer E (1976) Biochem Biophys Res Comm 69:893

60. Stoops JK, Awad ES, Arslain MJ, Gunsberg S, Wakil SJ, Oliver RM (1978) J Biol Chem 253:4464
61. Stoops JK, Wakil SJ (1980) Proc Natl Acad Sci, USA 77:4544
62. Stoops JK, Wakil SJ (1981) J Biol Chem 256:8364
63. Schweizer E, Piccinini F, Duba C, Gunter S, Ritter E, Lynen F (1970) Eur J Biochem 15:483
64. Oesterhelt D, Bauer H, Lynen F (1969) Proc Natl Acad Sci, USA 63:1377
65. Oesterhelt D, Bauer H, Kresze GB, Steber L, Lynen F (1977) Eur J Biochem 79:173
66. Schweizer R (1984) In: Fatty acid metabolism and its regulation. Elsevier, Amsterdam, p 59
67. Singh N, Wakil SJ, Stoops JK (1985) Biochem 24:6598
68. Singh N, Wakil SJ, Stoops JK (1985) Biochem Biophys Res Comm 131:786
69. Schuster H, Rautenstrauss B, Mittag M, Stratmann D, Schweizer E (1995) Eur J Biochem 228:417
70. Ziegenhorn J, Niedermeier R, Nüssler C, Lynen F (1972) Eur J Biochem 30:285
71. Engeser H, Hübner K, Straub J, Lynen F (1979) Eur J Biochem 101:407
72. Engeser H, Hübner K, Straub J, Lynen F (1979) Eur J Biochem 101:
73. Stoops JK, Singh N, Wakil SJ (1990) J Biol Chem 265:16971
74. Katiyar SS, Briedis AV, Porter JW (1974) Arch Biochem Biophys 162:412
75. Kresze GB, Steber L, Oesterhelt D, Lynen F (1977) Eur J Biochem 79:191
76. Stoops JK, Ross P, Arslanian MJ, Aune KC, Wakil SJ, Oliver RM (1979) J Biol Chem 254:3194
77. Stoops JK, Wakil SJ (1981) J Biol Chem 256:5128
78. Kitamoto T, Nishigai M, Sasaki T, Ikai A (1988) J Mol Biol 203:1823
79. Stoops JK, Wakil SJ (1982) J Biol Chem 257:3230
80. Stoops JK, Henry SJ, Wakil SJ (1983) J Biol Chem 258:12482
81. Wang YS, Tian WX, Hsu RY (1984) J Biol Chem 259:13644
82. Singh N, Wakil SJ, Stoops JK (1984) J Biol Chem 259:3605
83. Joshi AK, Smith S (1993) J Biol Chem 268:22508
84. Foster RJ, Poulose AJ, Bonsall RF, Kolattukudy PE (1985) J Biol Chem 260:2826
85. Yuan ZY, Hammes GG (1986) J Biol Chem 261:13643
86. Chang SI and Hammes GG (1989) Biochem 28:3781
87. Witkowski A, Naggert J, Wessa B, Smith S (1991) J Biol Chem 267:18488
88. Smith S (1981) Meth Enz 71C:188
89. Knudsen J, Grunnet I, Dils R (1981) Biochem J 186:287
90. Krätzschmar J, Krause M, Marahiel MA (1989) J Bacteriol 171:5422
91. Raibaud A, Zalacain M, Holt TG, Tizard R, Thompson CJ (1991) J Bacteriol 173:4454
92. McDaniel R, Ebert-Khosla S, Hopwood DA, Khosla C (1993) Science 262:1546
93. McDaniel R, Ebert-Khosla S, Hopwood DA, Khosla C (1993) J Am Chem Soc 115:11671
94. Bartel PL, Zhu CB, Lampel JS, Dosch DC, Conners NC, Strohl WR, Beale JM, Floss HG (1990) J Bact 172:4816
95. Fu H, Ebert-Khosla S, Hopwood DA, Khosla C (1994) J Am Chem Soc 116:4166
96. McDaniel R, Ebert-Khosla S, Fu H, Hopwood DA, Khosla C (1994) Proc Natl Acad Sci, USA 91:11542
97. McDaniel R, Ebert-Khosla S, Hopwood DA, Khosla C (1994) J Am Chem Soc 116:10855
98. B, Summers RG, Wendtpienkowski E, Hutchinson CR (1995) J Am Chem Soc 117:6811
99. McDaniel R, Hutchinson CR, Khosla C (1995) J Am Chem Soc 117:6805
100. Malpartida F, Hopwood DA (1984) Nature 309:462
101. Hallam SE, Malpartida F, Hopwood DA (1988) Gene 74:305
102. Yu TW (1995) PhD thesis, University of East Anglia
103. H, Hopwood DA, Khosla C (1994) Chem & Biol 1:205
104. McDaniel R, Ebert-Khosla S, Hopwood DA, Khosla C (1995) Nature 375:549
105. Summers RG, Wendt-Pienkowski E, Motamedi H, Hutchinson CR (1992) J Bacteriol 174:1810
106. Shen B, Hutchinson CR (1993) Science 262:1535
107. Carreras CW, Pieper R, Khosla C (1996) J Am Chem Soc 118:185

108. Shen B, Nakayama H, Hutchinson CR (1993) J Nat Prod 56:1288
109. Shen B, Hutchinson CR (1993) Biochem 32:
110. Shen B, Hutchinson CR (1993) Biochem 32:6656
111. Crosby J, Sherman DH, Bibb MJ, Revill WP, Hopwood DA, Simpson TJ (1995) Biochim Biophys Acta 1251:32
112. Shen B, Summers RG, Gramajo H, Bibb MJ, Hutchinson CR (1992) J Bact 174:3818
113. Lynen F, Tada M (1961) Angew Chem 73:513
114. Gaucher GM, Shepherd MG (1968) Biochem Biophys Res Comm 32:664
115. Dimroth P, Walter H, Lynen F (1971) Eur J Biochem 13:98
116. Martin JF, Demain AL (1978) In: The filamentous fungi. Edward Arnold, London, p 426
117. Pflügl G, Kallen J, Schirmer T, Jansonius JN, Zurini MGM, Walkinshaw MD (1993) Nature 49:91
118. Kobel H, Loosli HR, Voges R (1983) Experientia (Basel) 39:873
119. Offenzeller M, Su Z, Santer G, Moser H, Traber R, Memmert K, Schneider-Scherzer E (1993) J Biol Chem 268:26127
120. Spencer JB, Jordan PM (1992) Biochem J 288:839
121. Dimroth P, Walter H, Lynen F (1970) Eur J Biochem 13:98
122. Scott AI, Beadling LC, Georgopapadakou NH, Subbarayan CR (1974) Bioorg Chem 3:238
123. Vogel G, Lynen F (1976) Meth Enz 43:520
124. Dimroth P, Greull G, Seyfferet R, Lynen F (1972) Hoppe Syler's Z Physiol Chem 353:126
125. Dimroth P, Ringelmann E, Lynen F (1976) Eur J Biochem 68:591
126. Bhogal P, Shoolingin-Jordan P. Incorporation of novel substrates into 6-methylsalicylic acid and triacetic acid lactone analogues by 6-methyl salicylic acid synthase. In: Polyketides: chemistry, biochemistry, molecular genetics. 1996. Royal Society of Chemistry, Bristol, UK, p 19
127. Jordan PM, Spencer JB (1990) J Chem Soc, Chem Comm 238
128. Spencer JB, Jordan PM (1992) Biochemistry 31:9107
129. Woo ER, Fujii I, Ebizuka Y, Sankawa U, Floss H (1989) J Am Chem Soc 111:5498
130. Jordan PM, Spencer JB (1993) Biochem Soc Trans 21:222
131. Bedford DJ, Schweizer E, Hopwood DA, Khosla C (1995) J Bact 117:4544
132. Schröder J, Schröder G (1990) Z Naturforsch 45c:1
133. Hopwood DA, Sherman DH (1990) Annu Rev Genet 24:37
134. Lanz T, Tropf S, Marner FJ, Schröder J, Schröder G (1991) J Biol Chem 266:9971
135. Schöppner A, Kindl H (1984) J Biol Chem 259:6806
136. Tropf S, Kärcher B, Schröder G, Schröder J (1995) J Biol Chem 270:7922
137. Schüz R, Heller W, Hahlbrock K (1983) J Biol Chem 258:6730
138. Fliegmann J, Schröder G, Schanz S, Britsch L, Schröder J (1992) Plant Mol Biol 18:489
139. Woodward RB (1957) Angew Chem 69:50
140. Gerzon K, Flynn EH, Sigal MV, Wiley PF, Monahan R, Quarck UC (1956) J Am Chem Soc 78:6398
141. Griesebach H, Achenbach H, Hofheinz W (1960) Z Naturforschg 15b:560
142. Kaneda T, Butte JC, Taubman SB, Corcoran JW (1962) J Biol Chem 237:322
143. Cane DE, Hasler H, Taylor PB, Liang TC (1983) Tetrahedron 39:3449
144. Cane DE, Celmer WD, Westley JW (1983) J Am Chem Soc 105:3594
145. Cane DE, Liang TC, Taylor PB, Chang C, Yang CC (1986) J Am Chem Soc 108:4957
146. Cane DE, Yang CC (1987) J Am Chem Soc 109:1255
147. Cane DE, Ott WR (1988) J Am Chem Soc 110:4840
148. Caffrey P, Bevitt DJ, Staunton J, Leadlay PF (1992) FEBS Lett 304:225
149. Roberts GA, Staunton J, Leadlay PF (1993) Eur J Biochem 214:305
150. Marsden AFA, Caffrey P, Aparicio JF, Loughran MS, Staunton J, Leadley PF (1994) Science 263:378
151. Aparicio J, Caffery P, A.F.A. M, Staunton J, Leadley PF (1994) J Biol Chem 269:8524
152. Staunton J, Caffrey P, Apaicio JF, Roberts GA, Bethell SS, Leadlay PF (1996) Nature Struct Biol 3:188

153. Donadio S, McAlpine JB, Sheldon PJ, Jackson M, Katz L (1993) Proc Natl Acad Sci, USA 90:7119
154. Kao CM, Katz L, Khosla C (1994) Science 265:509
155. Kao CM, Luo G, Katz L, Cane DE, Khosla C (1994) J Am Chem Soc 116:11612
156. Cortes J, Wiesmann KEH, Roberts GA, Brown MJB, Staunton J, Leadlay PF (1995) Science 268:1487
157. Kao CM, Luo G, Katz L, Cane DE, Khosla C (1995) J Am Chem Soc 117:9105
158. Kao CM, Luo G, Katz L, Cane DE, Khosla C (1996) J Am Chem Soc, 118:9184
159. Pieper R, Luo G, Cane DE, Khosla C (1995) Nature 378:263
160. Wiesmann KEH, Cortés J, Brown MJB, Cutter AL, Staunton J, Leadlay PF (1995) Chem & Biol 2:583
161. Pieper R, Luo G, Cane DE, Khosla C (1995) J Am Chem Soc 117:11373
162. Luo G, Pieper R, Rose A, Khosla C, Cane DE (1996) Bioorg Med Chem 4:995–999
163. Pieper R, Ebert-Khosla S, Cane D, Khosla C (1996) Biochem 35:2054
164. Swan DG, Rodriguez AM, Vilches C, Mendez C, Salas JA (1994) Mol Gen Genet 242:358
165. Schupp T, Toupet C, Cluzel B, Neff S, Hill S, Beck JJ, Ligon JM (1995) J Bact 177:3673
166. Lipmann F (1980) Adv Microbiol Physiol 21:227
167. Kleinkauf H, von Doehren H (1990) Eur J Biochem 192:1
168. Zuber P, Nakano MM, Marahiel MA (1993) Am Soc Microbiol, Washington, DC, p 897
169. Kleinkauf H, v. Döhren H (1996) Eur J Biochem 236:335
170. Lipmann F (1954) In: The mechanisms of enzyme action. Johns Hopkins University, Baltimore, p 599
171. Laland SG, Froyshov O, Gilhuus-Moe CC, Zimmer TL (1972) Nature 239:43
172. Laland SG, Zimmer TL (1973) Essays Biochem 9:31
173. Lipmann F (1973) Acc Chem Res 6:361
174. Kleinkauf H, Roskoski R, Lipmann F (1971) Proc Natl Acad Sci 68:2069
175. Gilhuus-Moe CC, Kristensen T, Bredesen JE, Zimmer TL, Laland SG (1970) FEBS Lett 7:287
176. Diez B, Gutierrez S, Barredo JL, van Solingen P, van der Voort LHM, Martin JF (1990) J Biol Chem 265:16358
177. Smith DJ, Earl AJ, Turner G (1990) EMBO J 9:2743
178. Weckermann R, Fuerbass R, Marahiel MA (1988) Nucl Acids Res 16:11841
179. Stachelhaus T, Marahiel MA (1995) FEMS Microbiol Lett 125:3
180. Stachelhaus T, Marahiel MA (1995) J Biol Chem 11:6163
181. Haese A, Pieper R, von Ostrowski T, Zocher R (1994) J Mol Biol 243:116
182. Weber G, Schoergendorfer K, Schneider-Scherzer E, Leitner E (1994) Current Genetics 26:120
183. Gocht M, Marahiel MA (1994) J Bacteriol 176:2654
184. Heaton MP, Neuhaus FC (1994) J Bact 176:681
185. Haese A, Schubert M, Herrmann M, Zocher R (1993) Mol Microbiol 7:905
186. Turgay K, Krause M, Marahiel MA (1992) Molec Microbiol 6:529
187. Hamoen LW, Eshuis H, Jongbloed J, Venema G, van Sinderen D (1995) Mol Microbiol 15:55
188. Saito M, Hori K, Kurotsu T, Kaneda M, Saito Y (1995) J Biochem (Tokyo) 117:276
189. Tokita K, Hori K, Kurotsu T, Kanda M, Saito Y (1993) J Biochem (Tokyo) 114:522
190. Pavela-Vrancic M, Pfeifer E, van Liempt H, Schäfer H-J, von Döhren H, Kleinkauf H (1994) Biochem 33:6276
191. Schlumbohm W, Stein TU, C., Vater J, Krause M, Marahiel MA (1991) J Biol Chem 266:23135
192. Stein T, Vater J, Kruft V, Wittmann-Liebold B, Franke P, Panico M, McDowell R, Morris HR (1994) FEBS Lett 340:39
193. Hoffmann K, Schneider-Scherzer E, Kleinkauf H, Zocher R (1994) J Biol Chem 269:12710
194. Yamada M and Kurahashi K (1969) J Biochem 66:529
195. Kanda M, Hori K, Kurotsu T, Miura S, Yamada Y, Saito Y (1981) J Biochem (Tokyo) 90:765
196. Kanda M, Hori K, Kurotsu T, Miura S, Saito Y (1989) J Biochem (Tokyo) 105:653

197. Hori K, Saito F, Tokita K, Kurotsu T, Kanda M, Saito Y (1994) J Biochem (Tokyo) 116:1202
198. Stindl A, Keller U (1994) Biochemistry 33:9358
199. Billich A, Zocher R (1987) Biochemistry 26:8417
200. Pieper R, Haese A, Schroeder W, Zocher R (1995) Eur J Biochem 230:119
201. Cosmina P, Rodriguez F, deFerra F, Perego M, Venema G, van Sinderen D (1993) Molec Microbiol 8:821
202. Zocher R, Keller U, Kleinkauf H (1982) Biochem 21:43
203. Lawen A, Traber A, Geyl D, Zocher R, Kleinkauf H (1989) J Antibiot 42:1283
204. Plattner PA, Nager U, Boller A (1948) Helv Chim Acta 31:594
205. Traber R, Hofman H, Kobel H (1989) J Antibiot 25:1
206. Stachelhaus T, Schneider A, Marahiel MA (1995) Science 269:69
207. Fu H, Alvarez MA, Khosla C, Bailey JE (1996) Biochemistry 35:6527
208. Donadio S, Staver MJ, McAlpine JB, Swanson SJ, Katz L (1992) Gene 115:97

Angucyclines: Total Syntheses, New Structures, and Biosynthetic Studies of an Emerging New Class of Antibiotics

Karsten Krohn[1] · Jürgen Rohr[2]

[1] Fachbereich Chemie und Chemietechnik, Universität-GH Paderborn, Warburger Straße 100, 33098 Paderborn, Germany. *E-mail KK@chemie.Uni-Paderborn.de*
[2] Institut für Organische Chemie, Universität Göttingen, Tammannstr. 2, 37077 Göttingen, Germany. *E-mail: Jrohr@gwdg.de*

General and recent aspects of chemical syntheses on angucyclinones and selected related natural products are summarized here. In the first part of this review on angucyclines, the chemical reaction used as a key step for the construction of the angucyclinone frame is discussed. In the second part, new members of the angucycline group of antibiotics, their bioactivity and recent biosynthetic studies including modern aspects of molecular biology (genetics) are reviewed.

Table of Contents

Topics in Current Chemistry, Vol. 188
© Springer-Verlag Berlin Heidelberg 1997

1
Introduction

Resistance to antibiotics is an increasing problem and research on antibiotics
has seen a renaissance both in the private and the public sectors [1]. Resistance
against *Staphylococcus aureus* (hospitalism), killer bacteria, and the reoccurrence
of plague in India are only some of the problems to be addressed in the improve-
ment of known antibiotics and the search for new classes of antibiotics. An
emerging new group of antibiotics are the quinoide angucyclines [2, 3], which,
in contrast to the related linearly condensed anthracyclines, show not only anticancer activity but also a multitude of other interesting biological activities such
as antibacterial or antiviral properties and enzyme inhibition (reviewed by Rohr
and Thiericke [3]; for a review up to 1986 see [4]). Since 1991, approaches to the
chemical synthesis of the angucyclinones have become relatively standardized,
at least for the aromatic representatives, and many efficient methods have been
developed. Nonetheless, there are still a number of unsolved problems, particularly in the preparation of natural products which are nonaromatic in ring A
and/or B; these problems will be addressed here, in this first comprehensive
review on chemical synthesis. Finally, in the final sections, interesting data on
new angucycline structures, their biosyntheses, and their biological acitvity are
discussed.

In their review, Rohr and Thiericke classified the angucyclinones based on the
degree of oxygenation and C-glycoside formation (Scheme 1) [3].

This classification is also very useful for synthetic analysis, but there are additional structural features of importance for synthesis that will be demonstrated

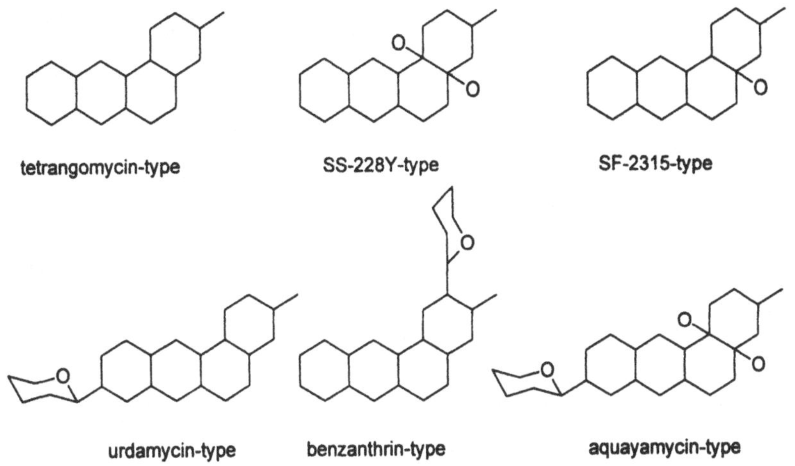

Scheme 1. Classification of angucyclines according to Rohr and Thiericke [3]

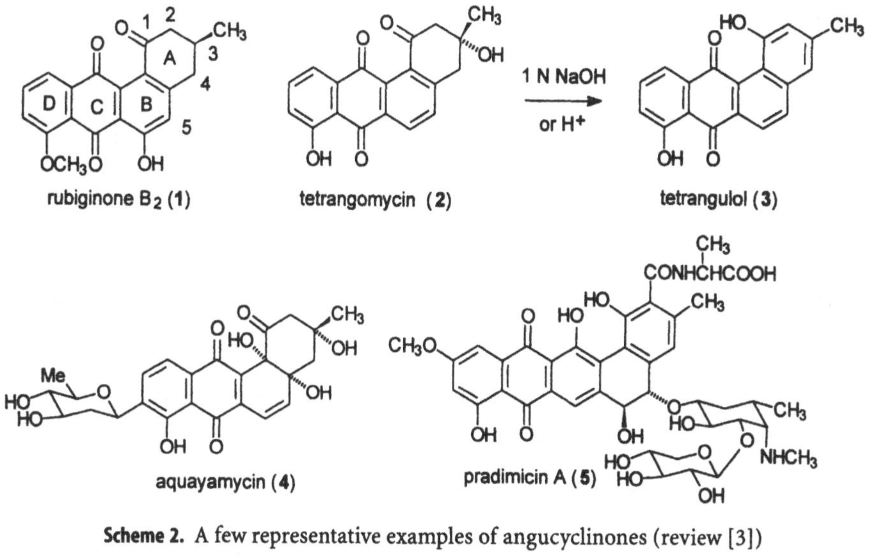

Scheme 2. A few representative examples of angucyclinones (review [3])

in Scheme 2 by some representative examples. Rubiginone B_2 (1) [5] is an example of a simple angucyclinone with an aromatic ring B but lacking the tertiary hydroxy group at C-3. This greatly facilitates synthesis. Despite the simple structure, the rubiginones have interesting vincristine cytotoxicity potentiating activity [6]. The first known examples of the benzo[a]anthraquinone antibiotics are represented by tetrangomycin (2) and tetrangulol (3), isolated by Kunstman and Mitscher in 1965 [7, 8]. Tetrangomycin (2) is missing the phenolic hydroxy group at C-6 but possesses a labile tertiary hydroxy group at C-3 that is easily eliminated by mild base or acid treatment to yield the simple 1,8-dihydroxy-benzo[a]anthraquinone named tetrangulol (3). Similar elimination products may be artefacts produced during isolation and purification rather than genuine natural products. The introduction of this unstable hydroxy group at C-3 was one of the first major obstacles to be overcome in angucyclinone synthesis. In aquayamycin (4), a tyrosine [9] and dopamine hydroxylase inhibitor [10], a sugar (olivose) is attached in *ortho*-position to the phenolic group at C-8. By definition [3] it is also a member of the angucyclinones since it has no hydrolyzable sugar. It represents the parent compound for a very large group of glycosidic angucyclines such as many urdamycines [2, 11, 12], P-1894B [13] (a prolyl hydroxylase inhibitor) or PI-083 [14] (a platelet aggregation inhibitor). Additional structural features are the two *cis*-hydroxy groups at C-4a and C-12b and the C-5,6 double bond. To date, the introduction of the tertiary hydroxy group at C-12b remains to be solved! Finally, the antibiotics with a benzo-[a]naphthacenequinone skeleton, represented by the important antifungal [15, 16] agent pradimicin [17], are also included, although they are not angucyclines in the strict sense and are not mentioned in the review of Rohr and Thiericke [3].

2
Nucleophilic Reactions

A key step in the synthesis of the simple aromatic bisphenol tetrangulol (3) by Brown and Thomson [18] was a Michael-type cyclization of a phenol to the chloronaphthoquinone moiety (Scheme 3). The starting material 8, connecting the naphthoquinone and the protected phenol, was prepared by an interesting radical alkylation of the chloronaphthoquinone 6 with a carboxylic acid 7 in the presence of silver ions and persulfate with concomitant decarboxylation (Torsell reaction [19]) to yield the dihydrobenzo[a]anthraquinone 9. The synthesis of tetrangulol (3) was concluded by dehydrogenation in boiling nitrobenzene.

Michael-type addition to naphthoquinones are common reactions but the anthraquinones as Michael acceptors are less well known. In connection with intramolecular addition of carbanions to anthraquinones such as 10 (compare Sutherland et al. [20], review [21]), we observed the formation of a naphthacenequinone 11 [22] (Scheme 4).

We wondered if this reaction might be exploited to construct the angularly condensed benzo[a]anthracenes. A problem was the aromatization of the initially formed hydroaromatic ring A under the relatively drastic basic reaction conditions. The starting material 12 was synthesized in a stepwise manner from 1-hydroxy-5-methoxy-9,10-anthraquinone [23]. The crucial cyclization can mechanistically be regarded as an intramolecular nucleophilic displacement of the methoxy group to afford a keto ester 13 with about 55% yield (Scheme 5). Only a few nucleophilic additions to electron-deficient anthraquinones are known [20, 24, 25] and intramolecular reactions of this type are more facile [21, 26–30]. The subsequent ethoxydecarbonylation under acidic conditions to yield ketone 14 presented no problem.

The 2-oxo-3-methylbenzo[a]anthraquinone (14) is an interesting and valuable angucycline intermediate. To demonstrate conversion into the racemic

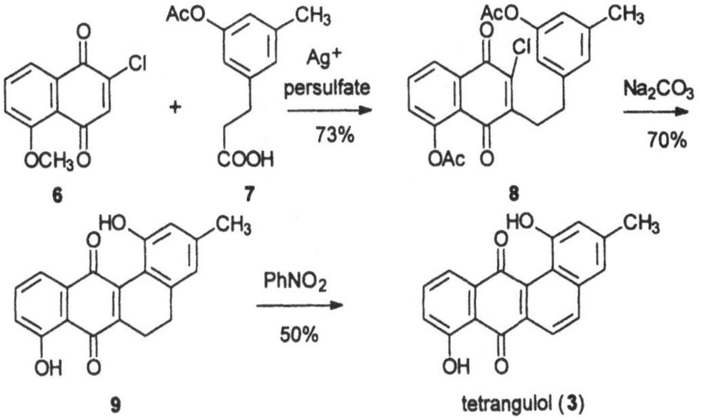

Scheme 3. Synthesis of tetrangulol (3) by Brown and Thomson [18] using a Michael reaction

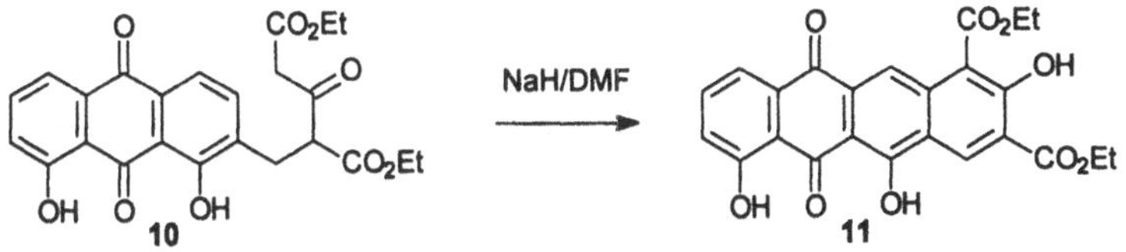

Scheme 4. Formation of naphthacenequinone 11 by intramolecular nucleophilic substitution [22]

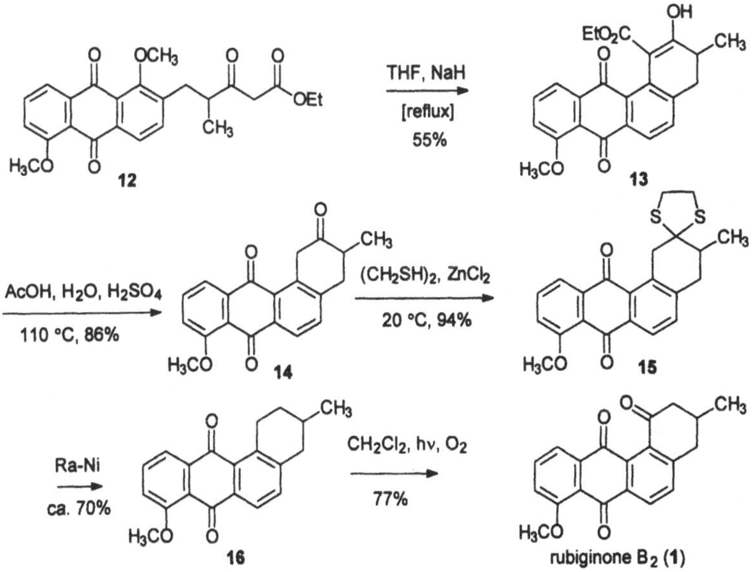

Scheme 5. Intramolecular nucleophilic addition to produce a dihydrobenzo[a]naphthoquinone (16) and transformation to rubiginone B$_2$ (1) [23]

natural product rubiginone B$_2$ (**1**), we effected a transposition of the carbonyl group from C-2 to C-1 (Scheme 5). Thioacetalization of **14** with 1,2-ethane dithiol generated the thioacetal (**15**) that was desulfurized to afford **16** by treatment with Raney nickel. The final transformation into (±)-rubiginone B$_2$ (**1**) was achieved by a novel photooxidation step (see Sect. 3.1).

The analogue of **1** hydroxylated at C-6, e.g. 3-deoxyrabelomycin (**20**), was synthesized by Kraus and Wu using the excellent Michael acceptor properties of the extremely electron-deficient 3-acetyl-5-methoxy-1,4-naphthoquinone (**17**) [31, 32] (for related reactions of Eugster et al. compare [33]). The addition of 5-methyl-1,3-cyclohexadione (**18**) proceeded at 20–25 °C without catalyst to afford adduct **19** after methylation. The subsequent intramolecular Michael-type reaction followed by elimination of methanol afforded more drastic conditions (NaOH, MeOH, 140 °C) and gave relatively low yields (27 %) of a tetracyclic intermediate which was demethylated in the usual manner with AlCl$_3$ to yield 3-deoxyrabelomycin (**20**, Scheme 6).

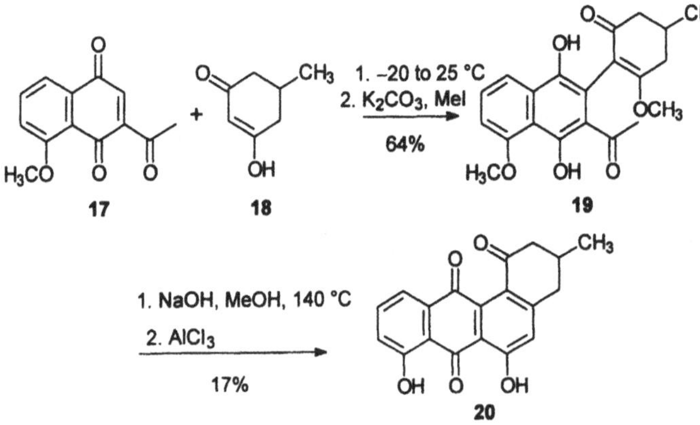

Scheme 6. Synthesis of 3-deoxyrabelomycin (**20**) using Michael additions [31, 32]

3
Electrophilic Substitutions

Electrophilic reactions on the electron-deficient anthraquinone are normally not possible. However, in 1936 Marschalk described the facile alkylation of the anthraquinone nucleus by aldehydes after reduction of the quinone to the electron-rich hydroquinone using dithionite [34]. This strategy might be called a "redox *Umpolung*", since the chemical reactivity of the anthracene core is reversed by the redox reaction.

In connection with a synthesis of 11-deoxydaunomycinone [35] we did not observe the expected linear cyclization of the bromide **21**, but rather the exclusive angular cyclization to the benzo[*a*]anthraquinone (**22**) [36]. This reaction was later exploited in the synthesis of a daunomycinone-rabelomycin hybrid **24** which incorporated structural elements of both antibiotics [37]. The required bromide **21** was prepared by alkylation of a bromomethylanthraquinone with acetylbutyrolactone followed by HBr treatment and acetalization. The anthrahydroquinone 21 underwent smooth cyclization to the angular system 22 (Scheme 7).

The final steps required the introduction of two oxygen atoms. The activated position α to the carbonyl group was brominated with copper bromide (compare with [38]) and solvolysis with aqueous sodium hydroxide afforded the *tert*-alcohol 23. The oxygen atom at C-1 was introduced by the photo-oxygenation step as mentioned above to yield the ketone 24. In fact, the photooxidation reaction on the deoxygenated precursor was discovered by serendipity – by leaving an NMR solution of the cyclization product on the bench in diffuse sunlight [37]. The hybrid structure showed considerable cytotoxicity in vitro without being a glycoside, a prerequisite in the anthracycline series.

The nucleophilic displacement on an anthraquinone core and also the electrophilic substitutions described above are limited to the synthesis of angucyclines with an aromatic ring B.

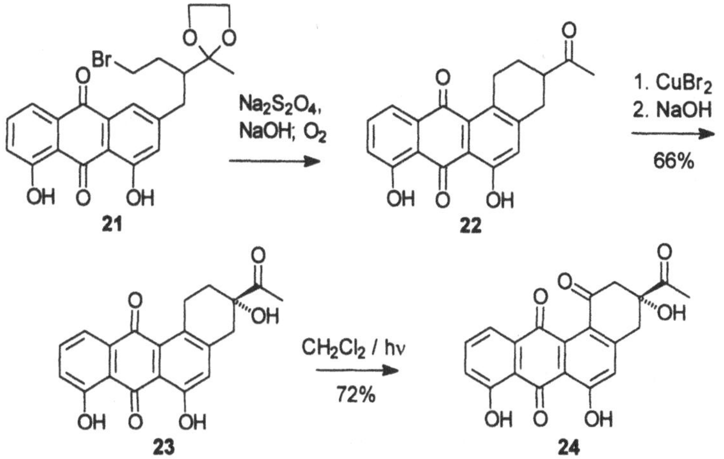

Scheme 7. Electrophilic cyclization of 21 to 22 followed by the introduction of an oxygen atom in the synthesis of a daunomycinone-rabelomycin hybrid 24 [37]

3.1
Photooxygenation at C-1

As mentioned above, the photooxidation was discovered by exposure of compound 22 to sunlight. The reaction proved to be of great value for angucycline synthesis because the β-hydroxy group present in most natural products which is easily eliminated under basic or acidic conditions (see Scheme 2) and the carbonyl group at C-1 can thus be introduced under mild neutral conditions. We assume that the reaction is initiated by Norrish type II γ-hydrogen abstraction of the excited carbonyl in 25 to yield a diradical 26 as shown in Scheme 8 with 1-deoxyrabelomycin (25) as the example [39]. The H-abstraction requires a very definite steric environment in which the benzylic protons have to be in proximity of the excited carbonyl group. Subsequent addition of the diradical 26 with singlet oxygen is supposed to yield a peroxodiradical **27** which can cyclize to an unstable hydroxy-1,2-dioxane (**28**). This intermediate is then opened up by proton abstraction to generate the C-1 carbonyl compound, e.g., rabelomycin (**29**). A mechanism involving triplet oxygen by Diels-Alder reaction with the mesomeric diene of the diradical structure **26** cannot be ruled out.

The noteworthy N-demethylation of aclacinomycin A [40] or the oxidation of 1-methylanthraquinone [41] probably proceeds by mechanistically related photooxidation processes [42]. For the photooxidation of natural rubiginones (OH at C-1) see [5].

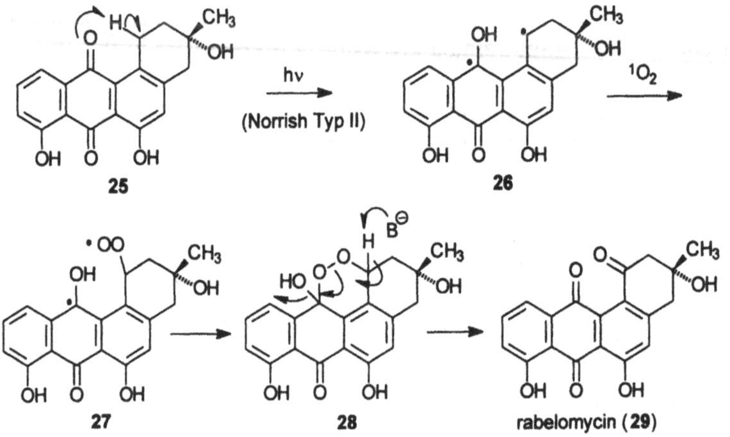

Scheme 8. Photooxygenation of 1-deoxyrabelomycin (**25**) to afford rabelomycin (**29**) [39]

4
Friedel-Crafts Reactions

The first synthesis of ochromycinone (**35**) [43] and its corresponding methyl ether X-14881 C (**34**) [44] was realized by Katsuura and Snieckus [45, 46]. The construction of the basic skeleton was performed by a Friedel-Crafts reaction of a carboxylic acid to yield an anthrone (**33**) that was subsequently oxidized to a dihydrobenzo[a]anthraquinone. The decisive step determining the regiochemistry was the coupling of the dianion formed by directed metalation of tetralol **30** with the benzaldehyde (**31**). The condensation product was not isolated but cyclized and dehydrated by treatment with acid to form the phthalide olefin **32** (77%). The carbonyl group at C-1 was introduced by a series of steps including selenohydroxylation and chromium(VI) oxidation followed by radical deselenation. Easier solutions such as the photooxygenation (see Scheme 8) or the base-catalyzed oxygenations for 3-deoxyangucyclinones [47] were later developed for introduction of the oxygen at C-1 (vide infra). The final demethylation of X-14881C (**34**) to the phenolic ochromycinone (**35**) using aluminum trichloride proceeded without problems (Scheme 9).

A related Friedel-Crafts reaction of a carboxylic acid derived from lactone **38** to yield an anthrone **39** was employed by Uemura et al. [48, 49]. However, a

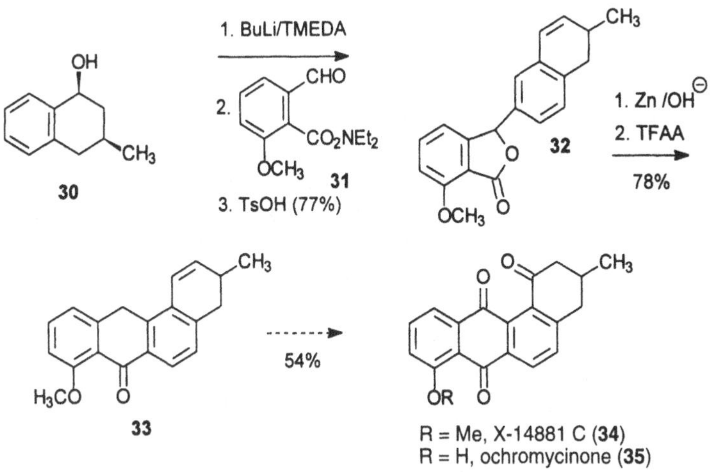

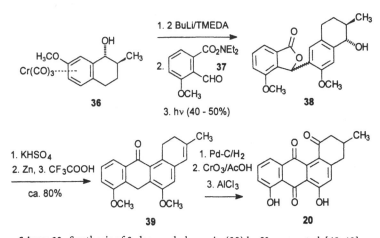

Scheme 9. Synthesis of ochromycinone and X-14881 C by Katsuura and Snieckus [45, 46]

(6-arene)chromium tricarbonyl complex **36** was used for the directed metalation. Subsequent coupling with the aldehyde **37** yielded lactone **38**. The base-catalyzed air oxidation of the unsaturated anthrone to the corresponding anthraquinone proved difficult, but chromium(VI) oxidation converted the hydrogenated derivative to the anthraquinone and simultaneously introduced the oxygen atom at C-1 to afford the racemic 3-deoxyrabelomycin (**20**) after cleavage of the methyl ethers with AlCl₃ (Scheme 10).

Scheme 10. Synthesis of 3-deoxyrabelomycin (**20**) by Uemura et al. [48, 49]

5
Diels-Alder Reactions

The Diels-Alder reaction is one of the most convenient ways to construct condensed six-membered ring systems. An early example of the construction of a hydrogenated benzo[a]anthraquinone **42** was given by Carothers and Coffman in 1932 by reaction of naphthoquinone **40** with a 2-chloro-vinyl-cyclohexene **41** [50] (Scheme 11). In fact, this kind of disconnection between the rings B and C thereby forming ring B is the most frequently applied in angucycline syntheses using the Diels-Alder reaction.

Related reactions of naphthoquinone (**40**) with styrene (**43**) as the diene component to yield the fully aromatized benzo[a]anthraquinone (**44**) are much more difficult and require long reaction times at elevated temperatures, as shown in Scheme 12 [51] (compare with [52]).

Guingant and Barreto [53] published the pioneering paper describing the synthesis of ochromycinone (**35**) by a Diels-Alder reaction. The dienone (**46**) was prepared from 3-ethoxy-5-methyl-cyclohex-2-enone (**45**) in two steps (alternatively, a diene with SPh instead of OMe could be used). The Diels-Alder reaction with juglone (**47**) was catalyzed with boron triacetate to overcome the somewhat poor reactivity of the electron-deficient diene **46**. The primary adduct **48** could not be isolated but directly eliminated and oxidized to ochromycinone (**35**) (Scheme 13).

The efficient procedure of Guingant and Barreto was used by Gould et al. [54] to prepare the aromatic tetrangulol (**3**) required for biosynthetic studies. As shown in Scheme 2, elimination of the tertiary hydroxy group of tetrangomycin (**2**) to form tetrangulol (**3**) is no problem. In the absence of this hydroxy group, conversion of ochromycinone (**35**) to the stereoisomeric mixture of the bromides **49/50** was required. Base-catalyzed quantitative elimination then produced the benzo[a]anthraquinone 3 (Scheme 14).

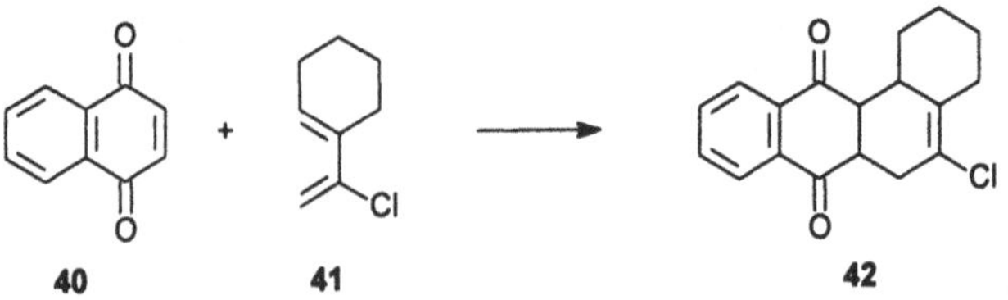

Scheme 11. Construction of a hydrogenated benzo[a]anthraquinone by Carothers and Coffman [50]

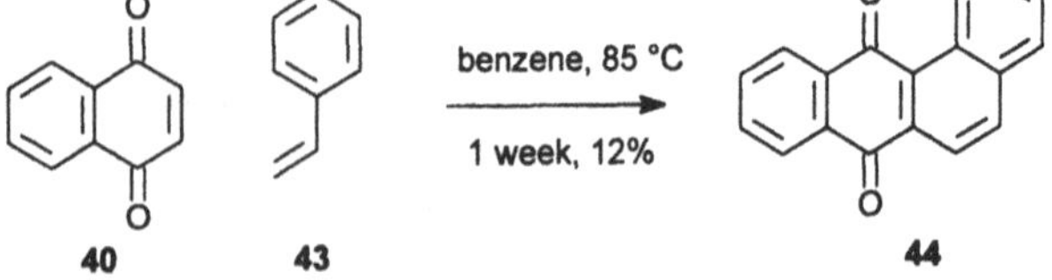

Scheme 12. Synthesis of benzo[a]anthraquinone by Diels-Alder reaction using styrene as the diene [51]

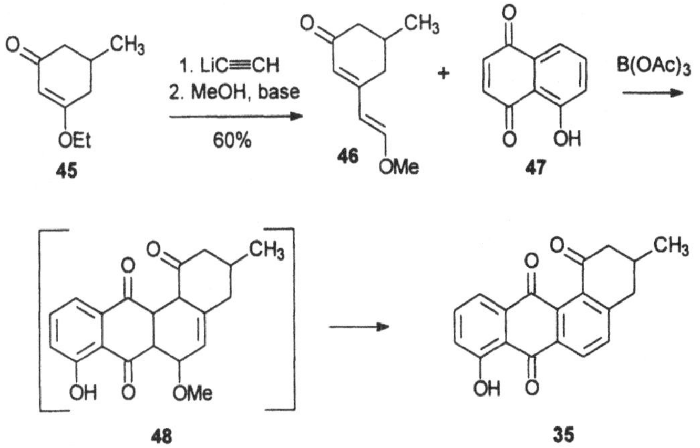

Scheme 13. Diels-Alder approach to ochromycinone (35) [53]

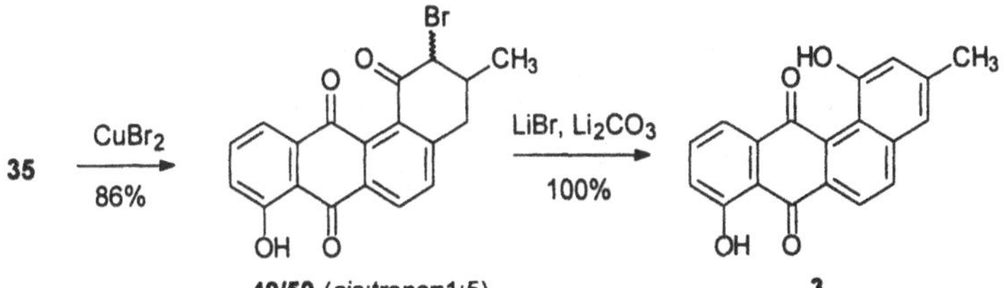

Scheme 14. Aromatization of ochromycinone (35) to tetrangulol (3) via bromides 49 [54]

In related investigations Valderrama et al. [47] studied the reaction of naphthoquinone (40) and juglone (47) with the ketene acetal 51. Using this oxidation state, the oxygen substituent in the product is always preserved at the terminal position of dienes (compare anthracycline chemistry [55]). Thus, the phenol ether 52a (61%) and smaller amounts of the phenol 52b were smoothly formed after silica gel-promoted elimination and air oxidation of the primary Diels-Alder adducts. The authors also observed base-catalyzed air oxidation to the ketone 53, analogous to similar oxidations of benzylic carbanions (compare with [56]).

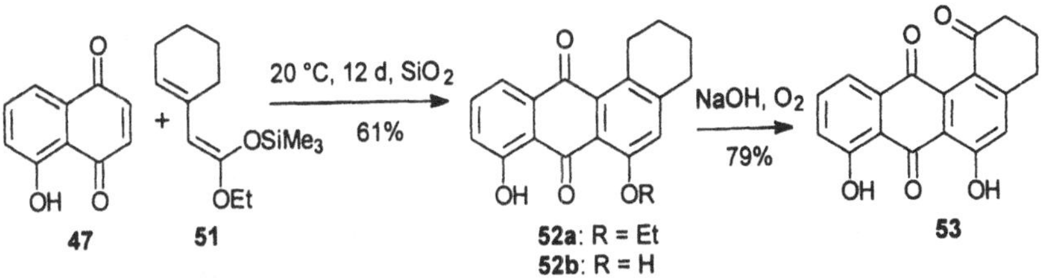

Scheme 15. Diels-Alder reaction using ketene acetal 51 and base-catalyzed air oxidation at C-1 [47]

The benzo[a]anthraquinone skeleton **56** was also formed in an interesting rearrangement of a spirodienone **55** prepared by Diels-Alder reaction of butadiene with the spiro-quinone **54** [57].

A different kind of disconnection (between rings B and C to construct ring C) was employed by Suzuki et al. [58, 59] (for related reactions compare with [60]). The arine **58** formed by treatment of iodotriflate **57** with *n*-BuLi reacted in a [2+4] cycloaddition with the siloxyfurane **59** to yield a mixture of the regioisomers **60** and **61** (1:14). Dehydrogenation of the major regioisomer **61** to the benzo[a]anthraquinone **62** with air in the presence of diazabicycloundecane (DBU) was quantitative (Scheme 16; for conversion to the glycoside C 104 see Sect. 8).

The synthetic scheme of Suzuki et al. [59] was based on the fact that adducts resulting from furan addition often rearrange to phenols (compare with [61]). In our group, we also investigated the reactions of naphthoquinones with siloxyfurans such as **64**. However, instead of the expected [4+2] cycloaddition, a Michael addition that proceeded without catalyst occurred in the reaction with 3-chlorojuglone (**63**) [39]. Interestingly, both the Michael acceptor **63** and the donor **64** reacted in a 1,4-reaction mode. The regio- and stereochemistry of the product **65** were confirmed by x-ray analysis (Scheme 18).

A similar diene **66**, prepared by reduction of the ketone **46** of Guingant and Barreto [53] (Scheme 13), was used by Larsen and O'Shea in the synthesis of the racemic rubiginone B$_1$ epimer (**69**) [62] and later (by a slight modification,

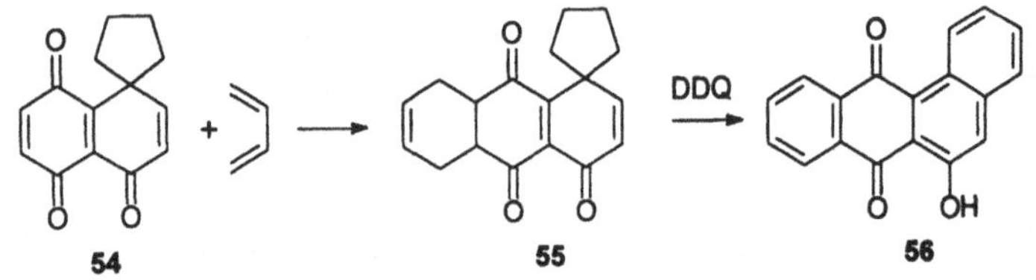

Scheme 16. Rearrangement of a spirodienone **55** to a benzo[a]anthraquinone [57]

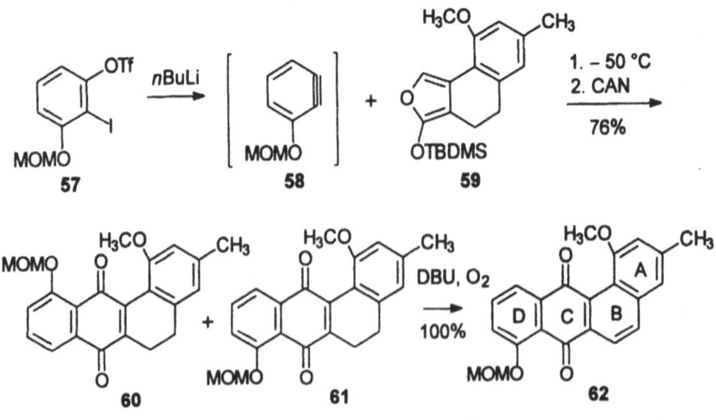

Scheme 17. Diels-Alder reaction of arine **58** with the siloxyfurane **59** [58, 59]

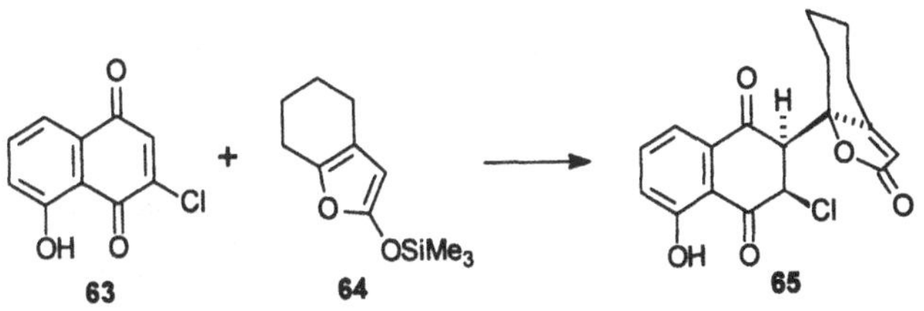

Scheme 18. Michael addition of the siloxyfurane 64 to the naphthoquinone 63 [39]

i.e., acetylation of 67) the racemic rubiginones [63] and emycin A [64]. The advantage offered by the diene 66 was that the adducts 67 with juglone 40 were relatively stable and did not spontaneously aromatize as observed by Guingant and Barreto [53]. Aromatization of 67 to 69 was induced by treatment with $[(AcO)_2B]_2O$ [63] or, in the case of the corresponding acetate, by DBU/air treatment [63, 64]. Oxidation of the hydroxy group at C-1 afforded the racemic rubiginone B_2 (2), and $AlCl_3$ demethylation ochromycinone (35) [64].

Interestingly, ring opening to the aldehyde 68 occurred on attempted aromatization of adduct 67 with either acid or base. Similar ring openings have also been observed with several natural products (e.g., aquayamycin 4 [65]) and establish the link to related open chain natural products such as the fridamycins and vineomycin B_2 [3].

A kinetic resolution with respect to the racemic diene 66 could be achieved in the Diels-Alder reaction promoted by a chiral Lewis acid prepared from BH_3 in the presence of (S)-3,3'-diphenyl-1,1'-binaphthalene-2,2'-diol and juglone (47) in the synthesis of (+)-emycin A and (+)-ochromycinone (35) [66] (compare previous investigations by Kelly [67]).

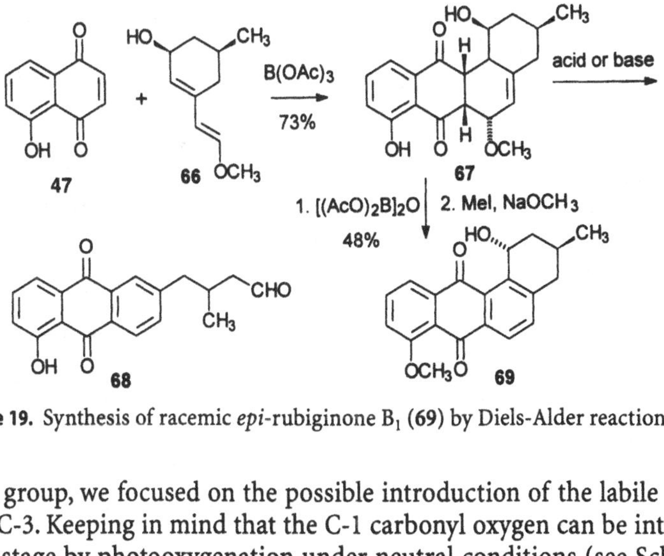

Scheme 19. Synthesis of racemic *epi*-rubiginone B_1 (69) by Diels-Alder reaction [62]

In our group, we focused on the possible introduction of the labile hydroxy group at C-3. Keeping in mind that the C-1 carbonyl oxygen can be introduced at a later stage by photooxygenation under neutral conditions (see Scheme 8), juglone (47) and the diene 70 resulted from the retrosynthetic analysis of the Diels-Alder route to rabelomycin (29).

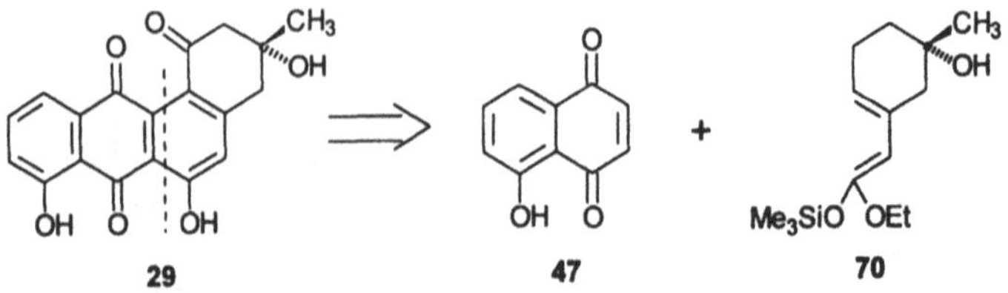

Scheme 20. Retrosynthetic analysis of rabelomycin (**29**) synthesis [68, 39]

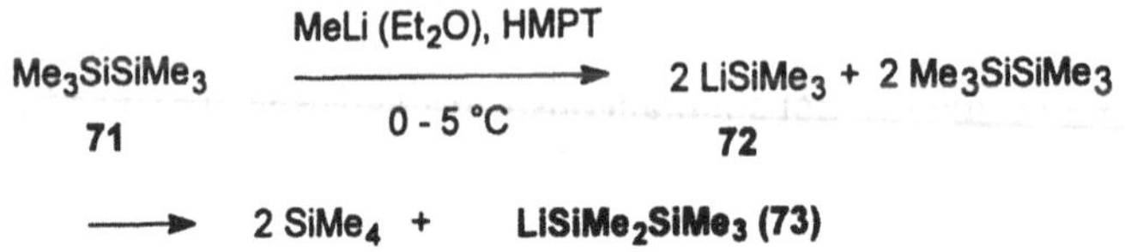

Scheme 21. Formation of lithium pentamethyldisilane (**73**) by reaction of Si$_2$Me$_6$ (**71**) with LiMe [68, 39]

We presumed that the required diene building block **70** would be relatively unstable and difficult to prepare. To solve the problem of the introduction of the labile tertiary hydroxy group, we decided to place a silicon atom at this position which can later be replaced by oxygen, as shown by the pioneering work of Fleming et al. [69]. The silicon serves two purposes: (1) it is a replacement for oxygen with a much decreased leaving group ability, (2) it directs regioselective deprotonation to generate the desired vinyl ketene acetal (see Scheme 24). Fleming proposed the dimethyl-phenylsilyl group as a precursor for the oxygen atom. Before testing this group we wanted to study introduction of the silyl group by use of the known trimethylsilyl lithium (**72**) prepared by Still by treatment of hexamethylsilyl lithium (**71**) with methyl lithium in HMPT [70]. However, when conducting the reaction on a much larger scale (molar instead of millimolar) we obtained the lithium pentamethyldisilane adduct **73** [68]. The formation of this species above 78 °C was observed earlier by Hudrlik et al. [71].

The conjugate addition of lithium pentamethyldisilane (**73**) to cyclic enones is a general reaction (e.g., **74** and **76**) and required no conversion to the corresponding copper species. The yields of the adducts **75** and **77** range from 50 % to 60 % [68].

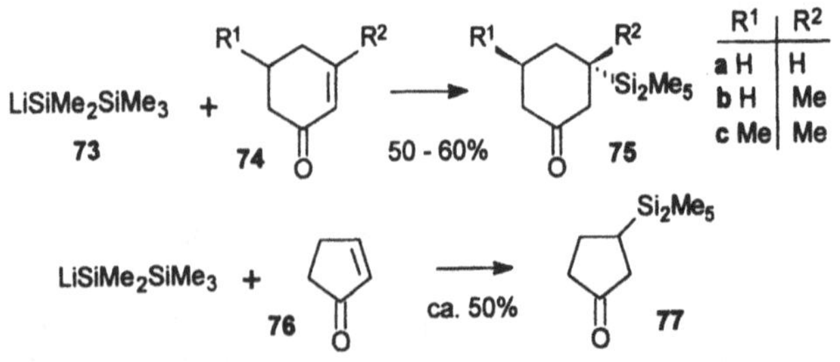

Scheme 22. Conjugate addition of lithium pentamethyldisilane (**73**) to cyclic enones [68]

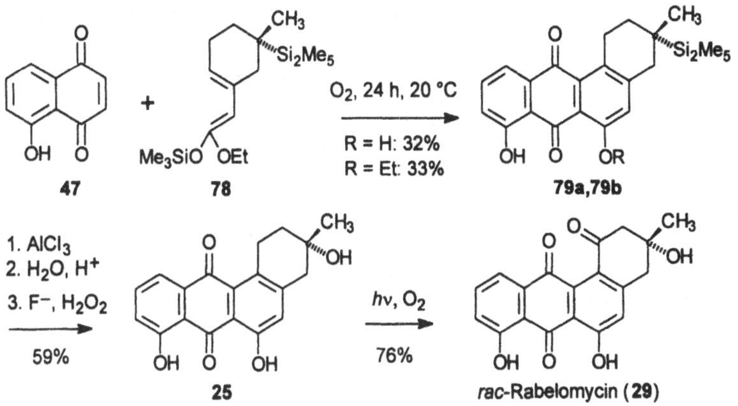

Scheme 23. Synthesis of *rac*-rabelomycin (**29**) using the silyldiene **78** in a Diels-Alder reaction [68]

Ketone **75 b** was converted to the corresponding vinyl ketene acetal **78** similarly as shown below (Scheme 24) for the dimethylphenylsilyl adduct **75 d**. Using this building block, the synthesis of racemic rabelomycin (**29**) was achieved as outlined in Scheme 23, including the conversion of the pentamethyldisilyl group of the adducts **79 a** and **79 b** into a hydroxy group by AlCl₃ cleavage of the Si-Si bond and cleavage of·the ethyl ether of **79 b** followed by H₂O₂ oxidation in the presence of fluoride [compare with 72]. The carbonyl group at C-1 was introduced by the photooxidation step of the 1-deoxyrabelomycin (**25**) described earlier (Scheme 8).

Later, the dimethylphenylsilyl adduct **75 d** was not only converted to the vinylketene acetal **81** by a Wittig-Horner reaction via the ester **80**, but also by reduction of **80** to the aldehyde **84** followed by silylation of the corresponding anion to the silylenol ether **85**. Vinylcyclohexenes without terminal substitution (e. g., **83**) was prepared via an alcohol derived from **80** or **84** [73, 74]. The least substituted diene **83** was alternatively prepared from the triflate **82** in a Stille coupling with trialkylvinyl stannane [75] (Scheme 24), a reaction also used by Toshima et al. [76].

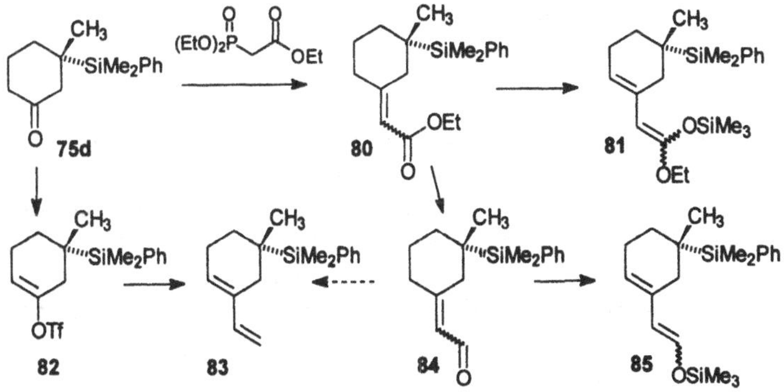

Scheme 24. Preparation of various vinylcyclohexenes from the silylketone **75 d** [73, 74]

With the required silyldienes **81**, **83**, and **85** in hand, the syntheses of most racemic angucyclines with aromatic ring B, such as rabelomycin **(29)** [39], tetrangomycin **(2)** and tetrangulol **(3)** [73] or the simple benzo[*a*]anthraquinone derivatives MM 47755 (isolation [77]) or X-14881 E (isolation [44]), could be prepared similarly as shown in Scheme 23 [74]. As a rule, the adducts using the ketene acetal **81** tend to rapidly eliminate one siloxy or alkoxy substituent generating a phenol or phenol ether. The allylsiloxanes derived from the enol ether **85** can also be aromatized, but the primary adducts are more stable. Finally, the primary Diels-Alder adducts derived from vinylcyclohexenes such as **83** are the most stable and perhaps also the most valuable for further conversion to non-aromatic angucyclines because the double bond can be oxygenated by various reagents (e.g., epoxidation, *cis*-hydroxylation, see below).

At this point it is instructive to examine the regiochemistry of the Diels-Alder reactions of the substituted naphthoquinones **47**, **87**, and **89** with the dienes **81** and **83** (Scheme 25) [74].

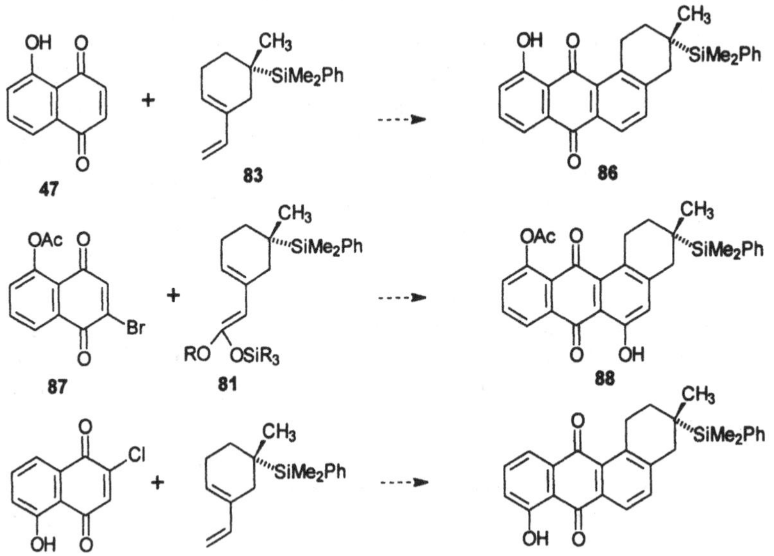

Scheme 25. Regioselectivity of some Diels-Alder reactions [74]

Ketene acetals such as **81** and juglone **(47)** form predominantly regioisomers derived from 1,8-dihydroxy-9,10-anthraquinone. This clearly follows previous experimental results [78, 79, 80] and theoretical frontier orbital considerations [81]. The regiochemical outcome results from an energetically favorable transition state in which the two larger orbital coefficients of dienophile LUMO and diene HOMO overlap. In ground state terms: the nucleophilic site of the diene adds to the more electrophilic site of the dienophile. The highly regioselective reactions of the oxygen-free and much less reactive diene **83** are less easily understood. Evidently, steric factors are also very important and in these cases addition of the less hindered terminal site of **83** adds to the less hindered site of the halogenated quinones **87** and **89** to form the adducts **88** and **90**, respectively [74].

For synthetic purposes it is important to take into account that elimination of hydrogen chloride or bromide from the primary Diels-Alder adducts starting from halogenated naphthoquinones is easy and the subsequent oxidative aromatization thereby facilitated. The synthesis of tetrangomycin (**2**) (Scheme 26) may demonstrate the principle, but many more synthetic examples of natural and non-natural angucyclines show the generality of regioselective Diels-Alder reactions as well as the utility of the silicon-oxygen exchange and the photooxidation reaction.

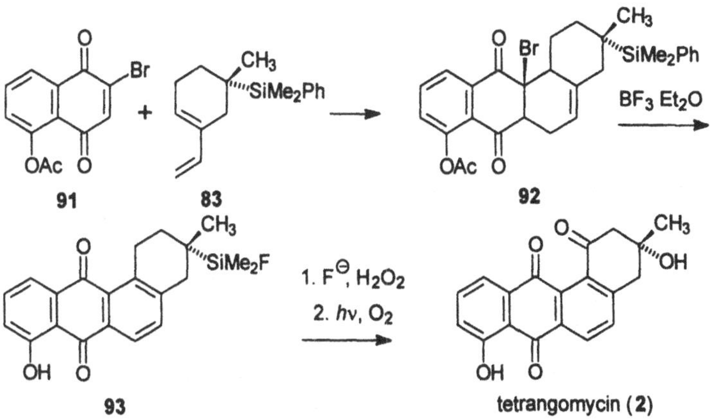

Scheme 26. Synthesis of tetrangomycin (**2**) by Diels-Alder reaction [73]

The Diels-Alder reaction is not only an excellent method for the synthesis of the aromatic angucyclines, but the double bond of the primary adduct (for example in **92**) is a functionality for the introduction of new oxygen atoms. The oxygen-free diene (**83**) is best suited for this purpose to prevent easy aromatization of ring B by elimination. However, halogenated naphthoquinones must be used in combination with this diene to control regiochemistry (see above). The relatively stable adduct **92** can be oxidized to the *cis*-diol **94** using catalytic amounts of osmium tetroxide and *N*-methylmorpholine-*N*-oxide (NMO) as the oxygen source (Scheme 27) [82]. Interestingly, good yields of the phenol **96** are obtained upon treatment of **94** with acid (elimination of both hydroxy groups was expected). However, a methoxy group is introduced at C-6 if **94** is treated with sodium methoxide in the presence of air to form **95**. This reactivity is explained via quinonemethides [82] that are also assumed in the reactions with molecular oxygen (Scheme 28) [83, 84] and NMO (see below Schemes 40, 41) [85, 86].

In model studies, Sulikowski et al. [83] investigated the reactivity of the Diels-Alder adduct **98** resulting from the reaction of bromonaphthoquinone **91** with the diene **97**. Two major types of oxidation products, the epoxides **100** and **101**, resulted from oxygenation in the presence of tetrabutylammonium fluoride

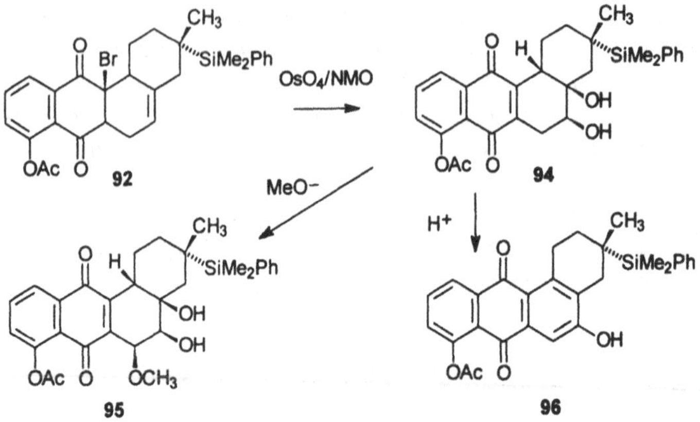

Scheme 27. Some reactions of the *cis*-diol **94** [82]

(TBAF) (the yields depend on the silyl protecting group at 1-OH). A quinone-methide radical is postulated to be the initial intermediate which is trapped by oxygen to give the hydroperoxy radical **99** [84]. The stereochemistry of the iso-lable oxidation products **100** and **101** depends on the stereoselective formation of only one stereoisomeric hydroperoxy radical **99**. The major compound **101**, with a triisopropylsilyl (TIPS) protecting group, was converted in a number of protection/deprotection and reduction steps into an analogue **102**, which has the stereochemistry of the SF 2315 B ring system [84].

Boyd and Sulikowski published the first synthesis of an enantiomerically pure diene building block **107** starting from (–)-quinic acid [87] in a sequence of reactions analogous to previous work by Steglich et al. [88] (Scheme 29). The epoxide resulting from tosylate **103** was reduced with LAH to the diol **104**. Mesyl-ation of the secondary hydroxy group followed by reductive fragmentation gave

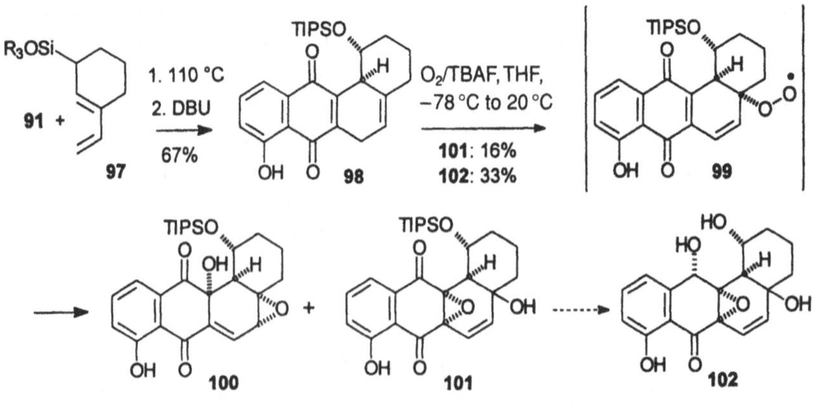

Scheme 28. Stereocontrolled assembly of the SF 2315 B ring system by air oxidation of adduct **98** [84]

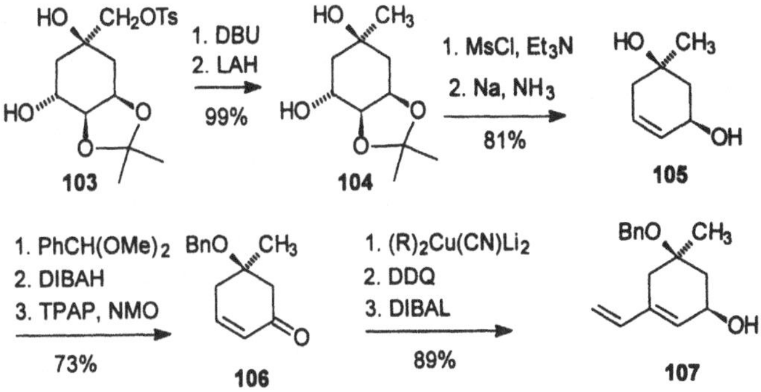

Scheme 29. Synthesis of an enantiomerically pure diene building block **107** starting from (−)-quinic acid [87]

the allylalcohol **105**. Protection of the more hindered tertiary alcohol of **105** was achieved by DIBAL reduction of the corresponding benzylidene acetal followed by oxidation with perruthenate to give ketone **106**. Conjugate addition of a higher order vinyl cuprate, trapping as silyl ether, and immediate DDQ oxidation regenerated the double bond. Finally, DIBAL reduction yielded the β-alcohol **107** as the major isomer (9:1).

The diene **107 a** was used in a total synthesis of the *C*-glycosidic urdamycinone B (**182**) and 104–2 (**183**) (see Sect. 8) but also in the synthesis and determination of absolute configuration of (+)-SF 2315 A (**111**) [89] (isolation of SF 2315 A see [90, 91]). Selective debenzylation of the Diels-Alder adduct derived from the quinone **91** and diene **107 a** by hydrogenation over palladium in the presence of potassium carbonate (which also effected dehydrobromination and saponification of the acetate) afforded the tertiary alcohol **108** (Scheme 30). The α-epoxide **109** was isolated exclusively upon epoxidation with dimethyldioxiran followed by acetylation. The allylalcohol **110** was generated by epoxide allylalcohol rearrangement using tetrabutylammonium fluoride (TBAF) as a mild base.

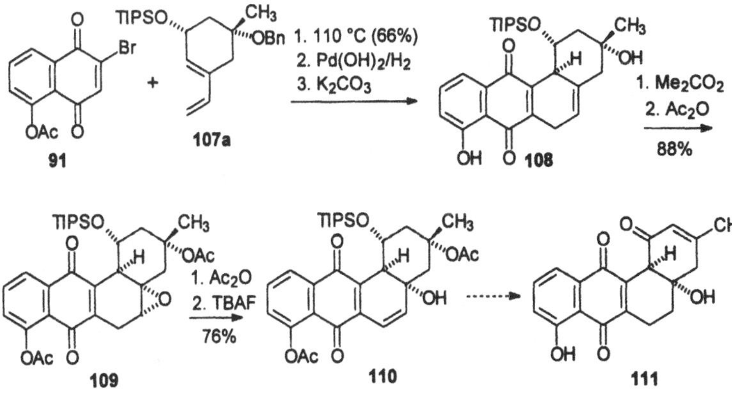

Scheme 30. Enantioselective synthesis of (+)-SF 2315 A (**111**) by Kim and Sulikowski [89]

A number of protections/deprotections and one oxidation step gave the angucyclinone 111 of the SF 2315 type (compare with [82]), establishing the absolute
configuration of the natural product.

Three successive [2+4] cycloadditions were used in the synthesis of the
pentacyclic methyl ether of G-2N by Kraus and Zhao [92] and later, by a slightly
modified procedure, also of the natural product G-2N (118) [93] (Scheme 31).
Thermal reaction of the cyclobutanol 112 with acrylic ester gave the dihydro-
naphthalene 113 which was demethylated by treatment with boron tribromide
and converted into the exocyclic ketene acetal 114. This unstable diene was reacted in a second cycloaddition with 2,6-dichlorobenzoquinone (115) to afford the
tetracyclic chloroquinone 116. In a last Diels-Alder reaction, ring E was anella-
ted by treatment of 116 with 1-methoxy-1,3-bis[(trimethylsilyl)oxy]-1,3-buta-
diene (117) to yield the pentacyclic natural product G-2N (118) [93].

Finally, Nicolas and Franck [94] attacked the problem of establishing the two
cis-hydroxy groups on an aromatic core (Scheme 32). As in many of their papers,

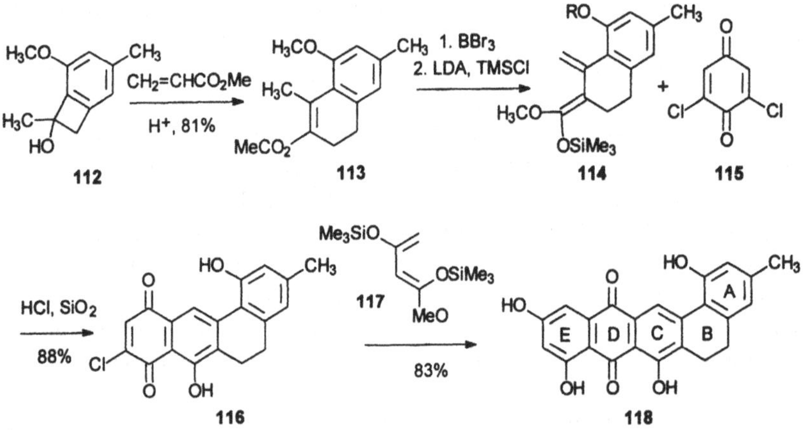

Scheme 31. Synthesis of G-2N (118) by three successive [2+4] cycloadditions [93].

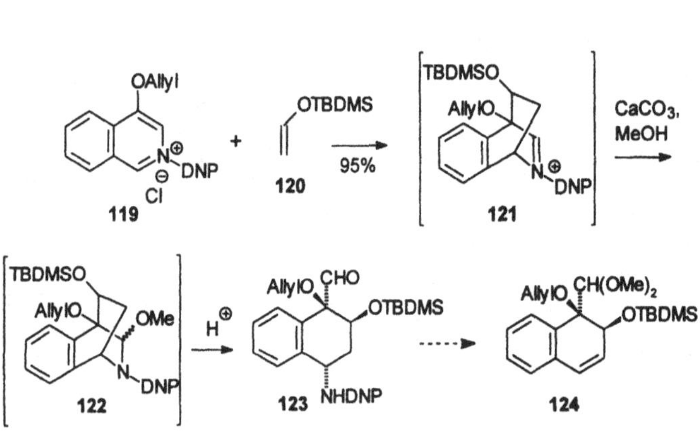

Scheme 32. Synthesis of the protected enediol 124 by Bradsher cycloaddition [94]

they used the Bradsher cycloaddition reaction to construct the ring system. The iminium salt 119 reacted with siloxyvinyl ether 120 to form the aldehyde 123 via the intermediates 121 and 122. Elimination of the amino substituent of 123 to form the protected enediol 124 was achieved after acetalization, saponification of the amide, and conversion to the methoiodide by exhaustive methylation.

6
Biomimetic-Type Syntheses

Biosynthesis is presumed to occur via a hypothetical decaketide (Scheme 33). Rohr has recently presented a folding analysis of polyketides in terms of sequential (E)- or (Z)-enolates [95].

However, it should be mentioned that Gould and Halley recently presented an alternative mechanism for PD 116198 involving a skeletal rearrangement [96]. We will later return to discussion of the problem of linear vs angular cyclization mode.

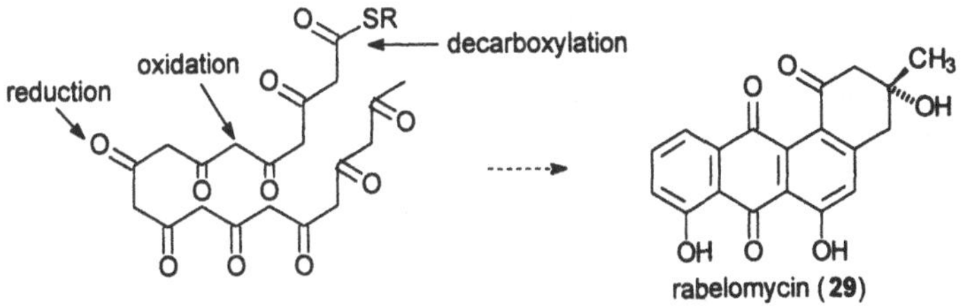

Scheme 33. Postulated decaketide cyclization to angucyclines [3]

The first biomimetic-type synthesis of (–)-urdamycinone B (131) (the enantiomer of the natural procuct) was realized by Yamaguchi et al. [97]. They initially examined the strategy by synthesis of racemic tetrangomycin (2) (R=H in Scheme 34). In particular, they achieved the irreversible dehydration/aromatization to tetrangulol (3) (compare with Scheme 2) but finally found mild conditions (NaOH, MeOH, 25 °C) to prevent the unwanted aromatization (77% for 2, R=H).

By a slight modification of the route, shown in Scheme 34, they also prepared the C-glycoside with L-olivose (132) (erroneously named rhamnose [97]) resulting in the enantiomer 131 of the natural product. Based on their earlier investigations of biomimetic-type condensations of dianions to aromatic systems [98], they prepared the naphthalenediol 126 by successive condensation of glutaric ester 125 with acetoacetate and acetate anions. The aldehyde 127 was then prepared by a sequence of palladium-catalyzed dealkoxydecarbonylation, base-catalyzed ring closure, protection as MOM ether, and DIBAL reduction. An important step was chain elongation of the aldehyde with lithiated acetylacetone monothioacetal. The intermediate diketoalcohol spontaneously cyclized with concomitant elimination and aromatization to 128. The dithioacetal protection

group prevented further uncontrolled cyclization steps. Acid-catalyzed deprotection, base-catalyzed air-oxidation to the anthraquinone, and removal of the
dithiane protection group then generated the crucial diketo intermediate 129.
Base-catalyzed cyclization afforded (–)-urdamycin B (131) (34%) together with
the diastereomer 130 (37%) (Scheme 34). No asymmetric induction of the
remote sugar could be observed on the newly formed chiral center at C-3.

It is noteworthy that – in principle – other cyclization products can also result
from the diketone 129. Evidently, conjugation to the aromatic ring stabilized the
enolate formed by deprotonation of the acetyl side chain at C-1 of 129 in such a
way that this anion reacted as the nucleophile in kinetically controlled aldol
reactions.

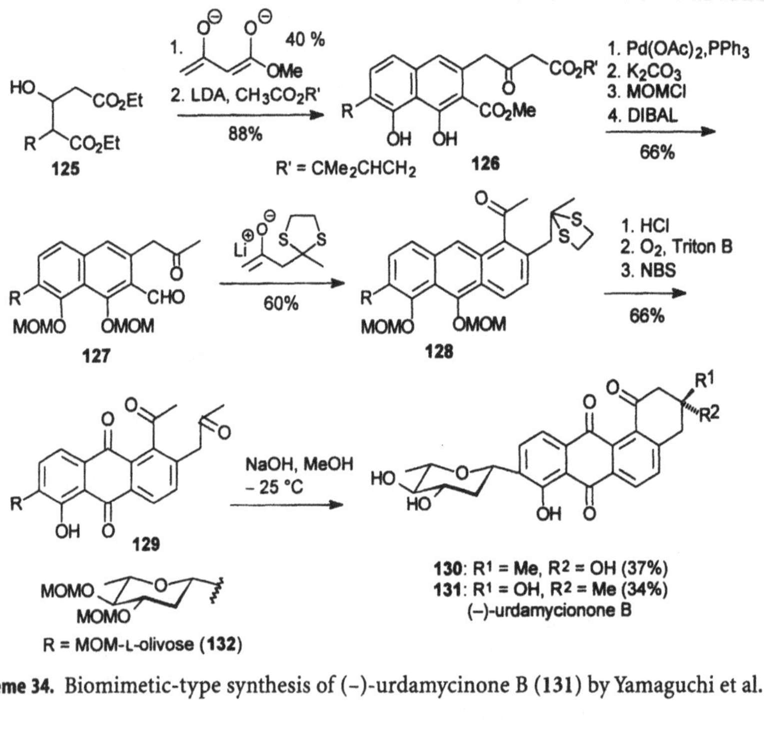

Scheme 34. Biomimetic-type synthesis of (–)-urdamycinone B (131) by Yamaguchi et al. [97]

In our work on biomimetic-type angucycline synthesis [86], we focused on
the control of cyclization mode and on the isolation of nonaromatized intermediates in the aldol cyclization of the oligoketide intermediates to arrive at
angucyclines of the SS-228Y- or SF-2315-type (see Scheme 1). The same decaketide precursor shown in Scheme 33 can potentially cyclize to the angular or
the linear framework, depending on the folding mode that is determined by the
enzyme. In chemical synthesis, the multitude of different condensations of the
polyketide has to be restricted. One efficient way to achieve this end is the
attachment of appropriate side chains at the *ortho*-position of ring systems, as
shown in the Yamaguchi synthesis (Scheme 34) and as described by Harris in
his review on biogenetic-type synthesis of polyketide metabolites [99]. In our

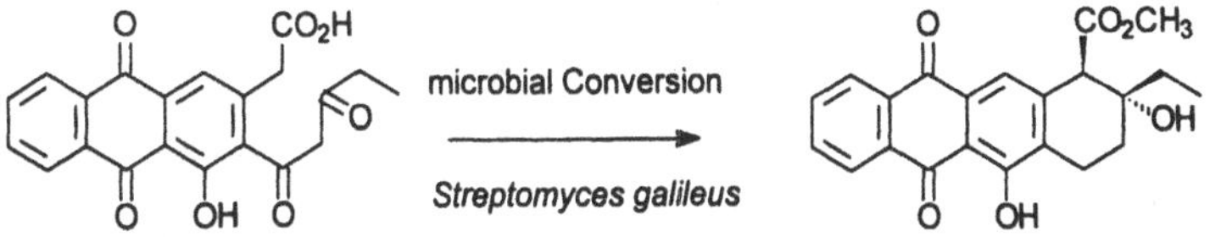

Scheme 35. Microbial conversion of 4-deoxyaklanonic acid (133) to 4,7-deoxy-aklavinone (133) [100]

investigation on microbial anthracycline synthesis via aklanonic acid derivatives [100 – 102], a linear cyclization mode was desired. An essential point was the presence of a carbonyl group at C-1 of the C5-side chain to modify the C-H-acidities in such a way that linear cyclization resulted in kinetically controlled aldol reactions, as shown in Scheme 35, demonstrating the microbial conversion of 4-deoxyaklanonic acid (133) to optically active 4,7-deoxyaklavinone (134).

The example shows that the nucleophilic centers have to be correctly opposed to the corresponding electrophilic carbonyl groups. To direct the cyclization into an angular mode, we decided to simplify the system by omitting the carbonyl group on the benzylic position. The naphthoquinone dibromide starting material 138 can easily be prepared from 1,5-dihydroxynaphthalene (135) followed by acetylation and bromination to 2-bromo-5-acetoxy-1,4-naphthoquinone (91) [103] (Scheme 36). Radical methylation with dimethylsulfoxide catalyzed by Fenton's reagent gave the methylation product 136 in 62 % yield [104]. Ester cleavage and methylation to 137 followed by NBS bromination then afforded the dibromide 138 [compare with 105]. The vinylic bromide serves two purposes: (1) it sterically prevents unwanted premature Michael addition of the ketoester to the quinone double bond and (2) it activates the vinylic position for attachment of the second side chain (see below).

The attachment of a C5-side chain by S_N2-alkylation at the intended benzylic position with the protected ketoester 139 as the nucleophile proceeded reliably and in good yield to form the alkylated ketoester 140 [86]. The second chain could also be coupled very efficiently by reacting the vinylic bromide 140 in a palladium-catalyzed Stille reaction [106] (review [107]) with the allyl tin compound 141 to afford the dialkylated naphthoquinone 142 (Scheme 36). Halonaphthoquinones have previously been coupled in a palladium/copper-catalyzed reaction by Echavarren et al. [108].

The reaction sequence was also conducted with other substituents on the naphthoquinone ring (e.g., OMe in 137 replaced by OH or H). Cleavage of the double bond in the side chain was the next step in our synthetic plan. Ozonolysis gave irreproducable results but the Lemieux-Johnson reaction $(OsO_4/NaIO_4)$ cleaved the double bond of the side chain without affecting the quinone. The ester 143 was a good model to study the cyclization mode. On reaction with Lemieux-Johnson reagent, a direct conversion into the linearly condensed system 144 was observed. Evidently, the highly acidic position of the β-ketoester immediately adds to the electrophilic carbonyl group generated by double bond cleavage in the opposite side chain (Scheme 37) [86]. Elimination

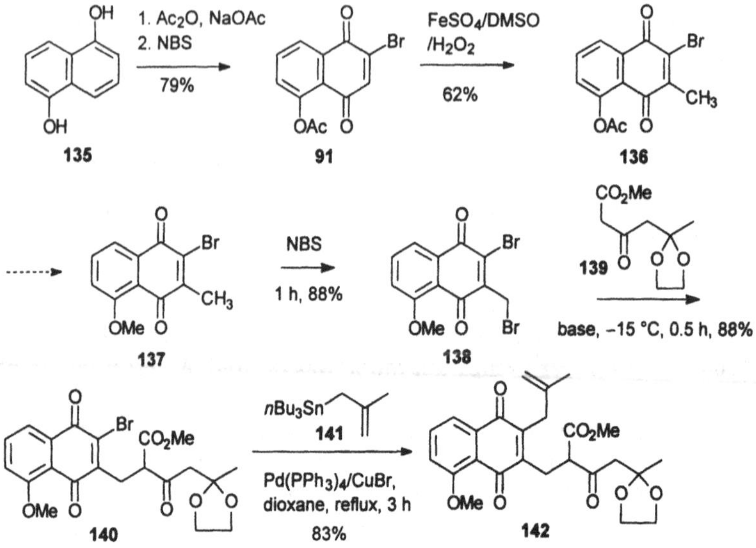

Scheme 36. Attachment of two side chains on the 1,4-naphthoquinone core [86]

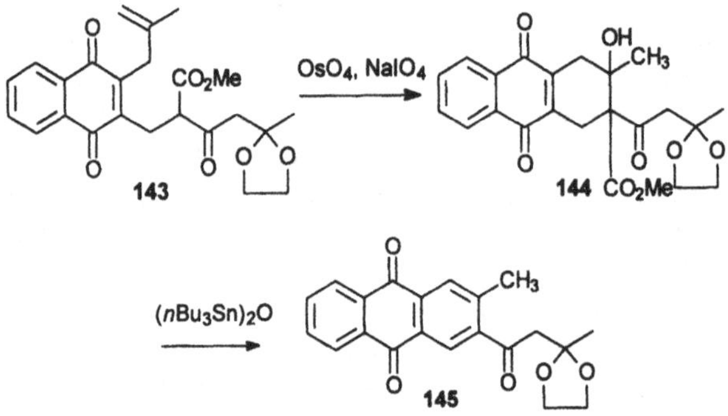

Scheme 37. Linear cyclization of ketoester 143 [86]

of water and methoxydecarbonylation was achieved by treatment of **144** with bis(*n*-tributyltin)oxide [109] to yield the decarboxylated derivative **145** of 4-deoxyaklanonic acid. This presents a potentially valuable route to anthracyclines without a hydroxy group at C-5 in ring B, which are otherwise not easily accessible [21, 55, 110].

Methoxydecarbonylation of **143** and related derivatives **146** (R=OMe, OH) under neutral conditions using bis(*n*-tributyltin)oxide [109] gave the protected ketone **147** which could be cleaved in the side chain to the diketone **148** in good yields and without premature cyclization. Subjection of **148** to treatment with a very mild base (K$_2$CO$_3$ in isopropanol) followed by clean cleavage of the ketal furnished the expected two stereoisomeric aldol products **149a** and **149b** (one in crystalline form) in which ring B of the angucycline is established and

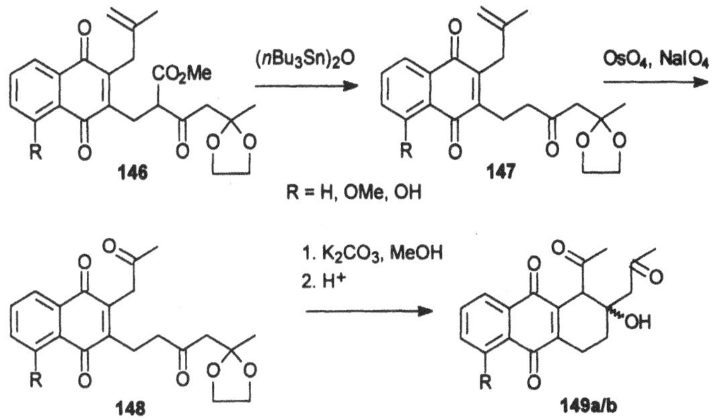

Scheme 38. Construction of ring B in the biomimetic-type angucycline synthesis [86]

the substituents are positioned to form the angularly condensed ring A (Scheme 38) [86].

With the primary cyclization products **149a/b** in hand, we next investigated the cyclization to form ring A of the angucyclines. In principle, aldol reaction of these intermediates should directly yield the products of the SF-2315 type (see Scheme 1). However, we were not yet able to cyclize these hydroaromatic precursors. Cyclization should be easier – as demonstrated by Yamaguchi et al. [97] – if the side chains are attached on a flat aromatic ring in which the reactive centers of the chains are forced to be much closer together in addition to the increased C-H-acidity of the acetyl side chain. Interestingly, in contrast to our expectation, the aldol products **149a/b** proved to be quite resistant to β-elimination. One explanation could be that enolate formation might be energetically disfavored due to steric hindrance and electronic repulsion of the opposed angular oxygens on the $sp2$ centers. Finally, NMO was tried as an oxidative reagent on **150a/b**. Not only the aromatization product **151** but also the corresponding phenols **152** were formed in the reaction (Scheme 39). The yield clearly

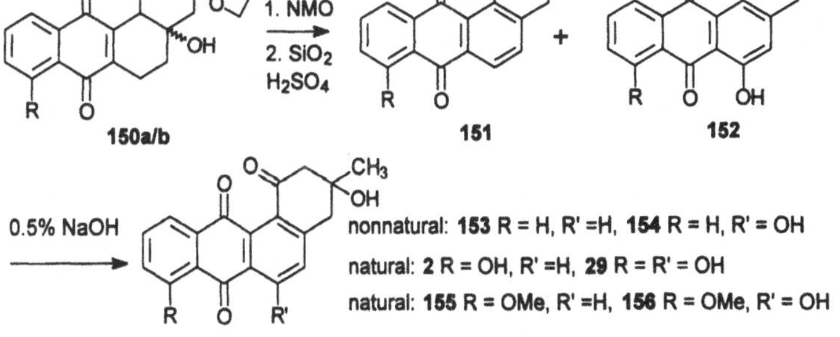

Scheme 39. Construction of ring A and hydroxylation at C-6 in biomimetic-type angucycline synthesis [86]

depended on the amount of NMO employed and up to 80% of the phenol **152** was isolated using 10 equivalents of NMO! The reagent was first used by Suli- kowski et al. [85] in this context and, in accordance with his proposal, we also assume quinonemethide tautomers as reactive intermediates for NMO addition (see Scheme 40).

The whole range of natural (**2, 29, 155, 156**) and also of the non-natural angucyclines **153** and **154** (not to mention their fully aromatized analogues) were now accessible by controlled aldol cyclization [86]. Our next goal will be the attachment of a C4 acetoacetic acid fragment onto the naphthoquinone core to arrive at partially cyclized precursors for biosynthetic studies similar to those performed with 4-deoxyaklanonic acid (**133**) [100].

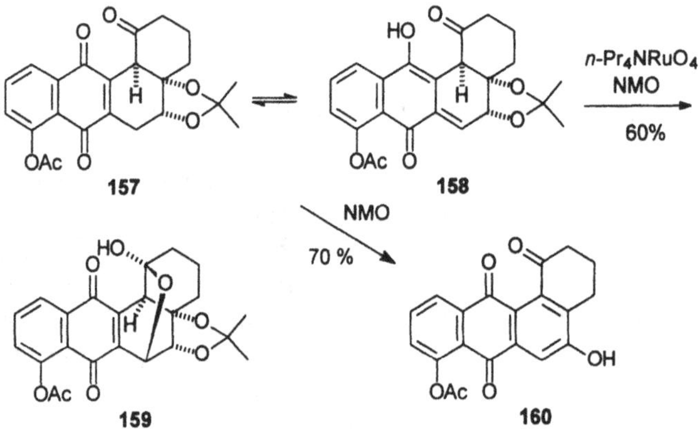

Scheme 40. Mechanism of NMO-oxidation of tetrahydroanthraquinones 157/158 [85]

As mentioned above, treatment of the aldol adducts **150a/b** with NMO pro- duced the phenol **152**. The interesting oxidation properties of NMO had pre- viously been investigated by Sulikowski et al. on the model compound **157** [85] (Scheme 40). They observed the formation of the hemiacetal **159** in 60% yield and assumed attack of the nucleophilic N-oxide on the quinonemethide tauto- mer **158** (or on the anion of **158**). A related reaction was observed in our group in which the diol **94** was methoxylated at C-6 to **95** by treatment with methoxide ions [82] (Scheme 27). An internal redox step is postulated to account for the reductive O-N-bond cleavage with concomitant oxidation of the hydroquinone back to the quinone. Without the presence of perruthenate, aromatization with formation of a C-5 phenolic hydroxy group was observed, a reaction later ex- ploited in the synthesis of angucycline 104–2 [87] (see Scheme 49). Thus, based on similar mechanistic principles, the chemical results of the NMO oxidations were quite different: compound **147** gave the C-6 phenol **152** [86] whereas **157/158** were converted to the C-5 phenol **160** [85].

7
Transition Metal-Mediated Reactions

The 1,4-naphthoquinone system is not only a good acceptor for stabilized carbanions, as exploited by Kraus and Wu [31, 32] (see Scheme 6), but also for radicals. The angular skeleton of the benzo[a]anthraquinone system is constructed in an elegant one-step transformation by a manganese (III)-induced radical addition of a malonic ester derivative 161 to 1,4-naphthoquinone 40 followed by addition of the newly generated radical to the benzene ring to form 162 (Scheme 41) [111]. However, the principle has not yet been used in advanced syntheses of angucyclinones.

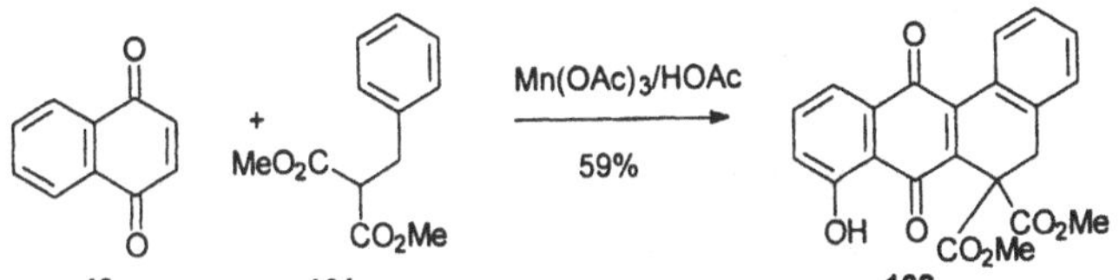

Scheme 41. Manganese (III)-induced radical addition to 1,4-naphthoquinone (40) [111]

Gordon and Danishefsky [112] used the reaction of a chromium Fischer carbene complex 164 with a cycloalkine 163 to build the naphthoquinone core 165 (Dötz reaction, review [113]), a procedure often used for synthesis of the linearly condensed anthracyclinones (e.g., [114]). The quinone ketone 165 has nucleophilic and electrophilic centers correctly positioned to furnish a benzo[a]anthraquinone. However, treatment with NaH or Triton B gave the spiro-compounds 166 as a mixture of two stereoisomers. These products evidently arose from Michael addition of the ketone enolate to the naphthoquinone double bond. But the weaker base DBU induced cyclization at ambient temperature to the benzo[a]anthraquinone 167 in 65% yield (Scheme 42). The primary aldol adduct apparently eliminated water and the resulting dihydrobenzo[a]anthraquinone aromatized under basic conditions in the presence of air. This is an instructive example of the influence of the base on the cyclization mode.

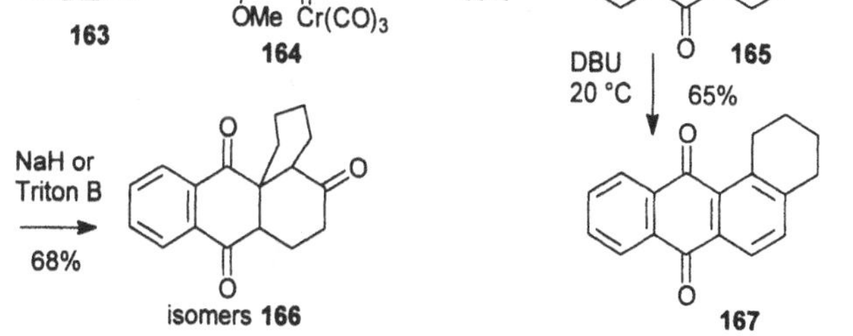

Scheme 42. Use of the Dötz reaction (CAN = ceric ammonium nitrate) in the synthesis of benzo[a]anthraquinone 167 [112]

Kelly and coworkers concentrated on the synthesis of analogues of the important pentacyclic antifungal agent pradimicin (5) (structure: [17, 115, 116], biological acitivity: [15, 117 – 120, review: 121]) and, e. g., G-2N and G-2A (171) (isolation: [122] revised structure: [123]). The tricyclic, chiral, B ring diol unit of pradimicin was prepared in model studies [124] and more recently the total synthesis of G-2N and G-2A (171) could be achieved [125] (Scheme 43). In a palladium-catalyzed Heck reaction, which required very elaborate special conditions, the vinyl-anthraquinone 168 was coupled with the iodide 169. The mixture of stereoisomeric coupling products were converted to the bistriflate 170 by reduction using Zn and concentrated HCl in N-methylpyrrolidine (NMP) (77%) followed by acid-catalyzed cleavage of the ketal and methylation of the resulting acid. The decisive intramolecular biaryl coupling was then accomplished in one operation through the agency of a palladium catalyst in the presence of hexamethyl tin for 7 days at 100 °C (44%). Brief exposure to an $AlCl_3/NaCl$ melt afforded G-2A (171), identical to an authentic sample [125]. The decarboxylation product G-2N was similarly obtained by fusion with pyridine hydrochloride.

In the work of Hauser and Carnigal [126], which led to revisal of the structures of G-2N and G-2A, condensation using a sulphonylphthalide and the enone 172 (see Scheme 44) followed by air oxidation of the primary adduct in DMF at 100 °C were used to construct the benzo[a]naphthacene-8,13-dione 173.

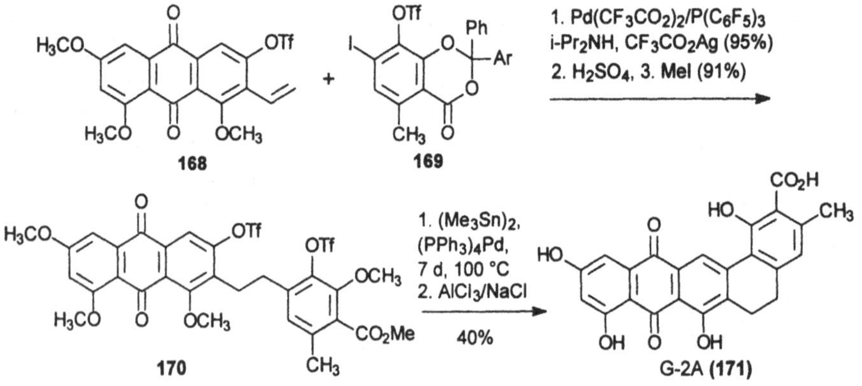

Scheme 43. Synthesis of G-2A (171) using Heck and Stille reactions [125]

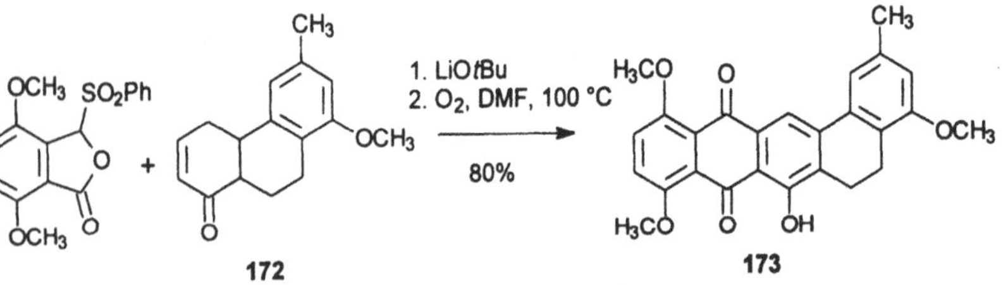

Scheme 44. Sulphonylphthalide condensation with enone 172 to construct a benzo[a]naphthacene-8,13-dione 173 [126]

8
C-Glycoside Syntheses

Many angucycline antibiotics possess *C*-glycosidically linked deoxysugars (olivose is usually the first sugar attached to the quinone) [3]. An example in which
the C-C bond to the sugar is introduced at a very early stage is shown in Scheme
34 in the biomimetic-type synthesis by Yamaguchi et al. [97] (for attachment of
the sugar see [127]). However, in most approaches to aromatic *C*-glycosides the
potential electrophilicity at the anomeric center of the sugar is used for the
coupling with the aromatic core which acts as the nucleophile (for reviews see
[128–130]. Direct electrophilic substitution of cationic sugar intermediates are
only possible with relatively electron-rich aromatic systems, as shown by Davis
[131, 132] and also Suzuki [133, 134] in the synthesis of the gilvocarcin group of
antibiotics. In some cases problems of regioselectivity (*ortho* vs *para* substitution) may occur. The nucleophilicity of the quinones is poor and the electrophilic substitution reaction with the cationic sugar species is usually conducted
with the reduced and protected hydroquinones, as exemplified by the vineomycin B_2 synthesis of Tius et al. [135, 136].

Naphthoquinone derivatives are excellent precursors for many types of
angucyclines, particularly in the Diels-Alder approach (see above). Consequently,
naphthols, which can easily be oxidized to naphthoquinones, have often been
selected as the starting materials for the attachment of the *C*-glycosidic part [137].

The nucleophilicity of the aromatic system can alternatively be increased by
using organometallic aromatic compounds, which also solves the problem of
regioselectivity. This method was applied by Sulikowski et al., who reacted a
variety of sugar lactones with aryl lithiums to afford intermediate lactols which
were subsequently reduced to the *C*-glycosides by cyanoboro hydride [137]. For
example, selective *ortho*-bromination of monobenzylated napthalene diol **174**
affords the bromide **175** which is converted into the dianion with three equivalents of *n*-BuLi. Treatment with the benzylated lactone **176** afforded lactol **177**
(Scheme 45, the D-dideoxygluconolactone **176** was erroneously drawn as the

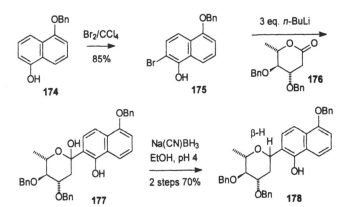

Scheme 45. *C*-glycoside synthesis by reaction of sugar lactones with lithium aryls [137]

L-enantiomer [137]). The final reduction step with cyanoboro hydride concluded the sequence to yield the C-arylglycoside 178. The corresponding C-aryl-
glycals can also be prepared by water elimination of the hemiacetal using the
Martin sulfurane (Ph$_2$S[OC(CF$_3$)$_2$Ph])$_2$) [138].

The C-glycoside 178 was used by Boyd and Sulikowski [87] in the total synthesis of enantiomerically pure urdamycinone B (182) and 104-2 (183) making use
of the diene 107 derived from shikimic acid (Scheme 29) and the NMO oxidation
to generate the C-5 phenols (Scheme 40). Thus, the bromonaphthoquinone 179
(prepared by treatment of phenol 178 with NBS) formed the tetracycle 180
through a Diels-Alder reaction with the diene 107 in analogy to sugar-free
reactants. Osmylation to a cis-diol, deprotection, oxidation, and acetalization gave
the acetonide 181. The decisive step in the aromatization to 182 and 183 was the
reaction with NMO (Scheme 46). Aromatization was also effected by direct
periodane oxidation of adduct 180 to derive 182 after deprotection.

A very direct synthesis of C-glycosides was published by Andrews and Larsen
[139], making use of the high nucleophilicity of 5,8-dimethoxy-L-naphthol (185)
in the Lewis acid-mediated reaction with acetylated 2-deoxy sugars such as tri-
acetyl olivose (184). Essential conditions for the success of the reaction were the
use of acetonitrile as the solvent and borontrifluoro etherate as the Lewis acid to
afford the C-glycosides in good yields (Scheme 47). The transformation to the
requisite naphthoquinones such as 186 then required only acetylation and
oxidation with cerium ammonium nitrate (CAN). Interestingly, only the products of ortho-substitution could be observed, which may be attributed to steric
hindrance of the peri position at C-4 by the neighboring methoxy group.

The synthesis of urdamycinone B (182) by Toshima et al. [76] is an instructive
example of the know-how that has been acquired in angucycline chemistry.
They directly coupled the unprotected olivose (188) with naphthol 187 using
trimethylsilyl triflate as the Lewis acid to afford the C-glycoside 189 after removal of the benzyl ethers and air oxidation in 21% yield. The strategy essentially
developed in our group was then employed in the Diels-Alder reaction of 189

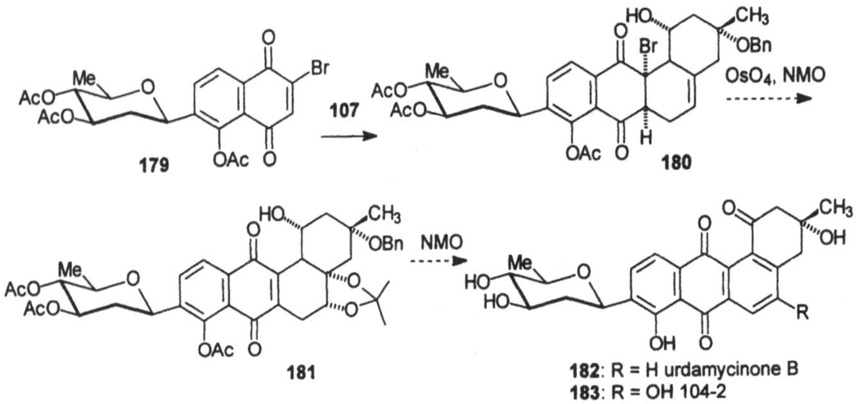

Scheme 46. Total synthesis of urdamycinone B (182) and 104-2 (183) [87]

with the silyl diene **190** to yield the adduct **191** after DBU-mediated elimination of thiophenol. Exchange of the silyl group by a hydroxy group and photooxygenation at C-1 were performed according to known procedures [39] (Scheme 48). As expected, no asymmetric induction (e.g., kinetic resolution) was observed in the Diels-Alder reaction of the enantiomerically pure *C*-glycoside **189** with the racemic diene **190** and separation of the diastereoisomeres was necessary.

The problem of selective *ortho-C*-glycoside formation was solved by Suzuki et al. [58] by Lewis acid-catalyzed (AgClO$_4$/hafnocene dichloride) *ortho*-selective rearrangement of primary *O*-glycosides to the corresponding *C*-glycosides. This strategy was already successfully employed in the synthesis of gilvocarcin group antibiotics [133, 134]. As outlined earlier, the electron-deficient quinones have to be reduced to electron-rich hydroquinones to enable the electrophilic attack of sugar species. Accordingly, the quinone **192** was reduced and converted to the substituted dimethoxynaphthalene **193**. *O*-glycosidation directly followed by rearrangement to the *C*-glycoside **195** was then mediated by the combined effects of hafnocene dichloride and silver perchlorate using the olivose derivative **194**. Deprotection, oxidation and selective protection of 3-OH of the olivose unit as a (*tert*-butyldimethyl)silyl ether then gave the monoalcohol **196** which was acylated and deprotected to yield the antibiotic C104 (**197**) [58] (Scheme 49).

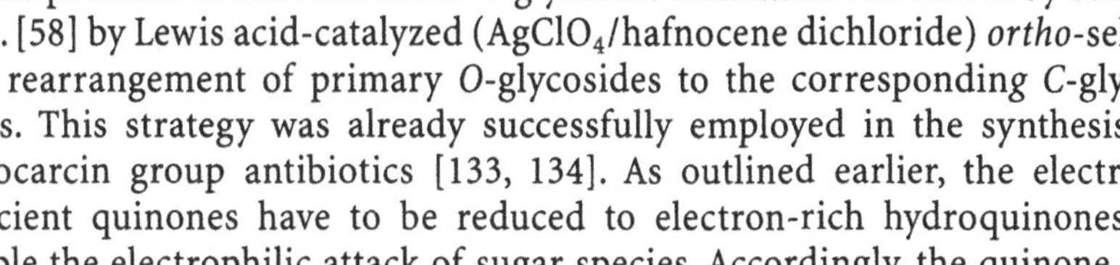

Scheme 47. Synthesis of *C*-glycosides by direct electrophilic substitution [139]

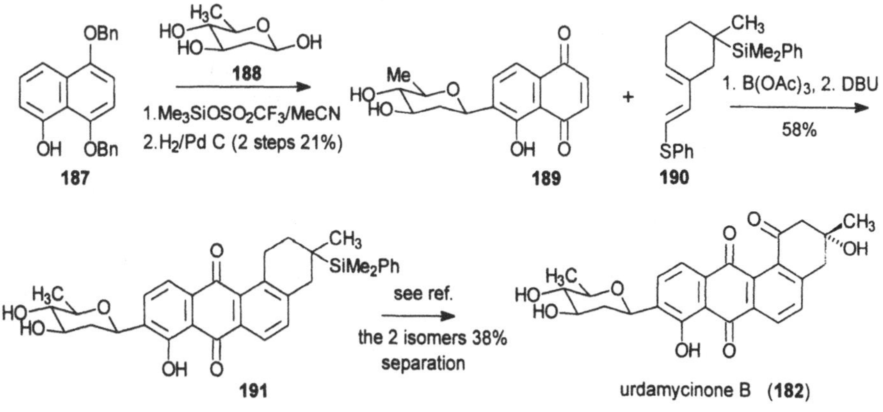

Scheme 48. Total synthesis of urdamycinone B (**182**) by Diels-Alter reaction [76]

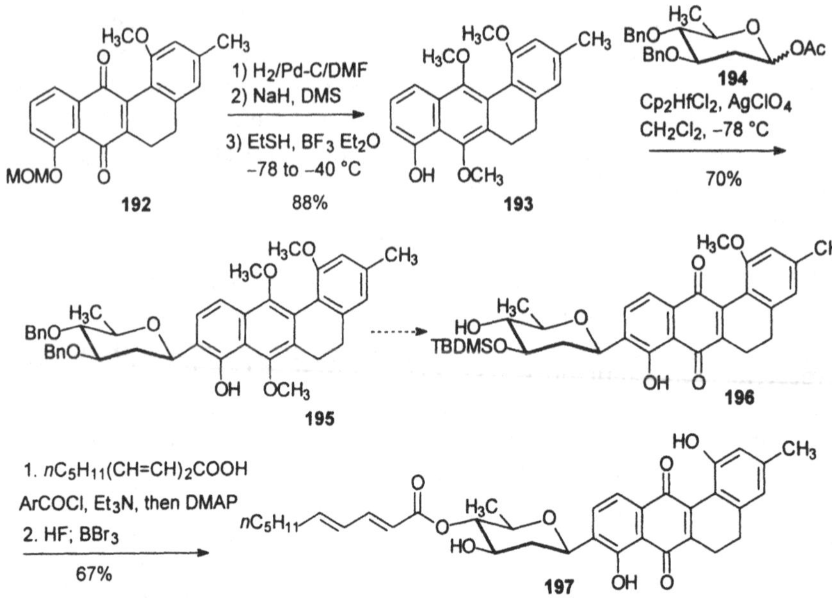

Scheme 49. Total synthesis of the antibiotic C104 by Suzuki et al. [58]

9
New Members of the Angucycline Group

The first review on angucyclines [3] contained structures published until the end of 1990. In this section, this overview is continued. Since this only requires reviewing the literature from 1991 until mid-1996, i.e. 5 years in contrast to about 25 years in the preceding review, the new compounds can be summarized here. Nevertheless, several new members of the angucycline group have been discovered during the last 5 years using various screening methods. Some unusual structures could be proven to derive biosynthetically from typical angucyclinones and thus have to be considered as members of this large group of decaketide-derived natural products. It seems more obvious than before that the typical molecular backbone of the angucyclinones, the tetracyclic frame derived from benzo[*a*]anthracene, plays a dominant role in biosynthetic routes of type II polyketides (see below and Sect. 10), since the angucycline group clearly outnumbers other major groups of tetracyclic decaketides, e.g., anthracyclines or tetracyclines.

To subdivide this section and facilitate an overview of the various new structures, a similar classification system, restricted to the new compounds, (Scheme 50) is used which is based on biosynthetic features leading to typical common structural elements of the aglyca, namely the *C*-glycosidic moiety and angular oxygens located at the AB ring fusion (C-4a/C-12b). This subdivision has turned out to be useful, even regarding synthetic approaches. If older angucyclines are mentioned in the text below without a given molecular formula and/or without literature citation, always refer to the previous review [3].

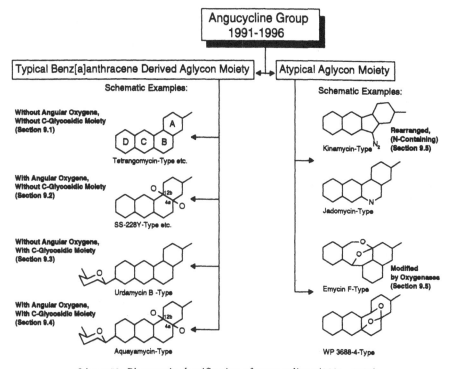

Scheme 50. Biogeneric classification of angucyclines (1991 – 1996)

9.1
Typical Angucyclinones and Angucyclines Omitting Both Angular Oxygens at the Ring AB Fusion and C-Glycosidic Moieties

This is the structurally most simple, but again the largest, subgroup of the newly discovered angucyclines.

Complete aromatic tetracyclic ring frames were found in chlorotetrangulol (**198**), WP 4669-brown (**199**) and BE-23254 (**200**, Table 1). The former two were discovered as minor components of the producer of the antitumor agent PD 116740, an unidentified actinomycete strain (WP4669). Both products may play a role in the biosynthesis of PD 116740; **199** has also been discussed as a possible artefact (decomposition product of **198**) [140]. The chlorine atom at C-13 suggests an unusual chloroacetate biosynthetic starter. BE-23254 (**200**), a *Streptomyces* sp. A 23254 product also containing chlorine, is an antitumor agent active against the human colon cancer DLD-1 (IC$_{50}$ 0.75 g ml^{-1}) [141]. Besides the chlorine atom at the 9-position, the absence of the methyl group at C-3, normally indicating the biosynthetic starter unit, is also unusual. Instead, the final carboxyl group of the biosynthesis is still linked at C-2. This COOH group usually gets lost during the biosynthesis through decarboxylation.

Several new, mostly very typical angucyclines or angucyclinones have been described which possess a partially saturated ring A and an anthra-quinone chromophore (rings B–D), the hatomarubigins A–D (**201–204**,

Table 1. Highly aromatic angucyclin(on)es of the tetrangomycin type

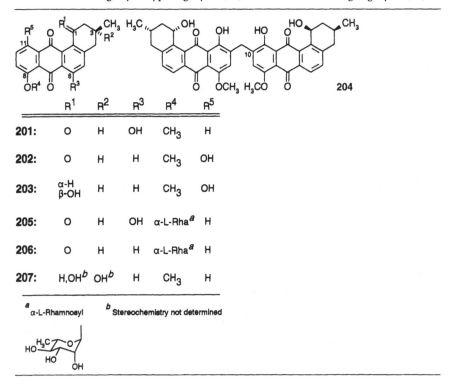

	R^1	R^2	R^3
198:	H	Cl	H
199:	H	OH	CH$_3$

Table 2. Novel tetrangomycin type angucyclin(on)es with saturated A-ring angucyclin(on)es

	R^1	R^2	R^3	R^4	R^5
201:	O	H	OH	CH$_3$	H
202:	O	H	H	CH$_3$	OH
203:	α-H β-OH	H	H	CH$_3$	OH
205:	O	H	OH	α-L-Rha[a]	H
206:	O	H	H	α-L-Rha[a]	H
207:	H,OH[b]	OH[b]	H	CH$_3$	H

[a] α-L-Rhamnosyl   [b] Stereochemistry not determined

CE 33 A-D), the atramycins A (205, BY 90-OH) and B (206, BY 90-H), and 34–1 (207, Table 2).

The hatomarubigins, isolated from *Streptomyces* sp. 2238-SVT4, are anti-tumor antibiotics which enhance the cytotoxicity of colchicine against multi-drug-resistant KB cancer cells (human squamous cell carcinoma) [142, 143]. Structures 202–204 show a rare 11-OH group: hatomarubigin C (203) is a reduced variant of hatomarubigin B (**202**). Most unique is the dimeric structure of hatomarubigin D (**204**). This is not due to a common phenol coupling, since the two halves are linked by a methylene bridge (at C-10), which may originate biogenetically from the C1-pool (e.g., formaldehyde).

The atramycins A (**205**) and B (**206**), from *Streptomyces atratus* BY 90 [144, 145], contain an interesting α-phenolglycosidically connected L-rhamnose moiety. The aglyca of **205** and **206** are the known angucyclinones 3-deoxyrabelomycin (**20**) and ochromycinone (**35**), respectively. Phenol glycosides among natural products are quite exceptional anyway; among the angucyclines they were only found once in the landomycins (see below). The atramycins are active against P388 mouse leukemia.

Antibiotic 34–1 (**207**) is the most recent of several known antibiotics (34–2: X-14881 E=8-O-methyl-3, 34–3: 6-deoxy-8-O-methylrabelomycin **155**, 34–4: 8-O-methylrabelomycin **156**). All are active against gram-positive bacteria and were isolated from *Streptomyces fradiae* strain 34, which is a construct obtained by intraspecific protoplast fusion of two *S. fradiae* strains. The parent strains were known as producers of the aminoglycoside antibiotic neomycin and the macrolide antibiotic tylosin, respectively [146]. Because of its reduced C-1 carbonyl, **207** resembles emycin A and hatomarubigin C (**203**), its closest relative.

Several members of the angucycline subgroup described in this section possess saturated carbons in ring B or C, in contrast to the structures described above. This is preferably due to redox reactions of their biosyntheses (Tables 3, 4; see also Sect. 10).

The structures of the landomycins produced by *Streptomyces cyanogenus* S-136 have been revised due to extensive biosynthetic studies on landomycin A (**208**, see also below) [147]. As a consequence the deoxysaccharide chains are linked to 8-OH and not at 11-OH, as previously reported [148]. Structure **208** exhibits remarkable antitumor activities, especially against prostate cancer cell lines. The bioactivity of the landomycins is very likely due to an interference with DNA, which is strongest for landomycin A, the molecule with the longest deoxysaccharide chain [149].

TAN-1085 (**209**) [150], produced by the unspecified *Streptomyces* sp. strain S-11106, is described as an inhibitor of both angiogenesis and aromatase. As in the landomycins, ring B is partially saturated. The sugar moiety is indicated as being rhodinose.

A completely reduced and thus saturated ring B is observed in ochracenomycin B (**214**) [151], the major new compound found in the culture broth of *Amicolatopsis* sp. MJ 950-89 F4. No biological activity has been yet discovered for **214**. The two other new ochracenomycins are biologically and structurally more interesting, both possessing angular oxygens (see Sect. 9.2.).

Antibiotic-683 (**210**) [152] reveals an unusual OH residue at C-4 and is therefore related to elmycin E and X-14881 D [3]. The nearly completely saturated ring C is probably a result of biosynthetic redox reactions leading to an epoxide ring between C-6a and C-12a and a reduced quinone system (see also elmycin C). Structure **210** is produced by *Streptomyces* sp. strain Y-83,30683 and shows antibacterial and extraordinary anticancer (mouse leukemia L-1210, IC_{50} 3 g ml^{-1}) activities. Another recent relative of **210** and elmycin C is angucyclinone D **211**,

Table 3. Tetrangomycin type angucyclin(on)es with patially reduced A-, B-, or C-rings

	R¹	R²	R³	R⁴	R⁵
208:	H	H	H	DHSC[a]	OH
209:	OH	OH	Rho[b]	H	H

[a] Deoxyhexasaccharide chain

$(\alpha$-L-Rhodinosyl-1-3-β-D-Olivosyl-1-4-β-D-Olivosyl)$_2$

Landomycins B, C, D = Variants regarding the deoxysaccharide chain
(see also section 10)

[b] Rhodinosyl (absolute stereochemistry and glycosidic bondage not determined)

Table 4. Tetrangomycin type angucyclin(on)es with patially reduced A-, B-, or C-rings

	R¹	R²	R⁴	R³, R⁵
210:[a]	H	OH	H,OH	O (Oxirane)
212:[a]	OH	H	H,OH	— (Double Bond)
211:[b]	OH	H	O	O (Oxirane)

[a] Stereochemistry of the entire molecule is not determined

[b] Stereochemistry of the entire molecule is relative

from *Streptomyces* sp. strain Gö 40/14, which also has a 6a,12a-epoxide ring at the fusion of rings B and C. The relative stereochemistry of the molecule has been determined through x-ray analysis [153]. No antibacterial and antifungal activities have been detected yet.

Hydranthomycin (**212**) [154], a new agroactive antibiotic, is a close structural (redox) isomer of SM 196 A [3]. It differs from the latter only in its secondary alcohol function at C-12 and the keto function at C-1 (SM 196 A: vice versa). Structure **212**, a product of *Streptomyces* sp. K93-5305, was discovered in a screening for herbicidal antibiotics using the plant *E. gracilis* as test organism. It also shows a moderate antifungal activity against *Pyricularia oryzae*.

Emycin C (**215**) and the oxygen-rich emycin H (**216**) are novel minor congeners of a blocked mutant of the emycin producer *Streptomyces cellulusae* ssp. *griseoincarnatus* (Scheme 51) (mutant 1114–2; see also Sects. 9.5 and 10.2) [155–157].

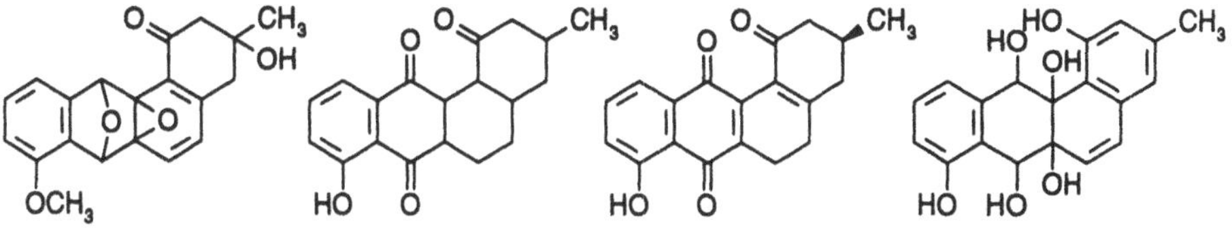

| 213 | 214 | 215 | 216 |

Scheme 51. Further angucyclinones of the tetrangomycin type with unusual oxygenations or reductions

The most unusual ring C-modified variant among the recent tetrangomycin (2)-type angucyclinones is SCH 58450 (**213**), a farnesyl protein transferase inhibitor (IC_{50} 29 M) from *Streptomyces* sp. SCH 58450. It exhibits an unusual diepoxide ring system [158a] and shows 25-fold selectivity for farnesyl transferase over geranylgeranyl protein transferase-1 and may be suitable as anti-Ras drug. Activated forms of Ras are associated with a variety of human cancers, with farnesylation being an essential step for Ras-induced cellular transformations. Based on results of model syntheses [158b], the structure is in doubt; thus, the alternative structure with the 6a,7-/12,12a-diepoxide structure should also be considered [158].

9.2
Typical Angucyclinones and Angucyclines with Angular Oxygen(s) at the Ring AB Fusion Omitting a C-Glycosidic Moiety

These compounds include the SS-228Y type, the SF-2315 type, the azicemicin type and the WP 3688-3 type (Scheme 52). Some new analogues of the SS-228Y-type have been discovered (Tables 5–8). The one with the simplest structure is the 2,3 dihydro analogue of SS-228Y, ochracenomicin A (**217**) [151]. This minor congener of the antibiotic complex mentioned above shows by far the best antibacterial (gram-positive/gram-negative) and fungicidal activity among the ochracenomycins – which may be due to the angular oxygens. Nonetheless, structure-activity relationships remain speculative as long as the stereochemistry of the chiral centers has not been determined.

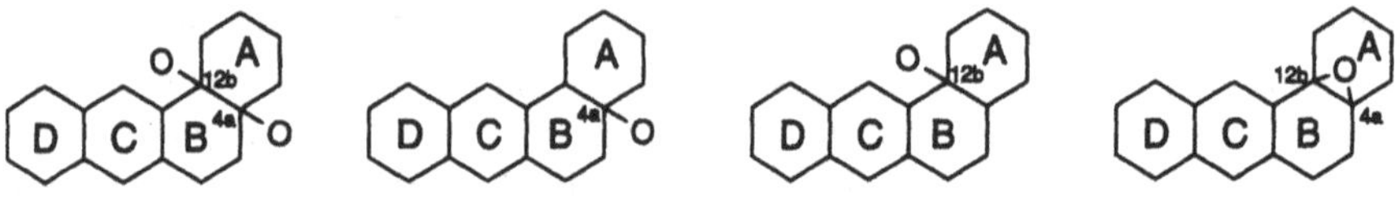

SS-228Y-Type   SF 2315-Type   Azicemicin-Type   WP 3688-3-Type

Scheme 52. Types of angucyclinones with angular oxygens

Table 5. Novel angucyclin(on)es with angular oxygens at the ring AB fusion (SS-228Y type)

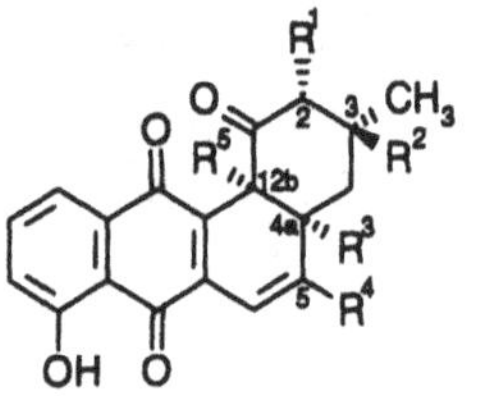

	R^1	R^2	R^3	R^4	R^5
217:	H	H[a]	OH[a]	H	OH[a]
218:	OH	OH	OH	H	OH
222:	OH[a]	OH[a]	O-D-Glu[b] O-Thioglu[c]	O-D-Rho[a,d]	
223:	OH[a]	OH[a]	O-D-Glu[b] O-Thioglu[c]	OH[a]	

[a] Stereochemistry not determined

[b] O-α-D-Glucosyl

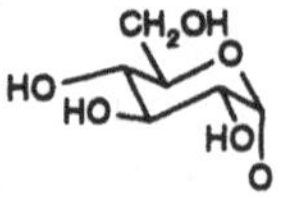

[c] O-α-2-Deoxy-2-mercaptoglucosyl-, absolute stereochemistry of the sugar moiety is not determined

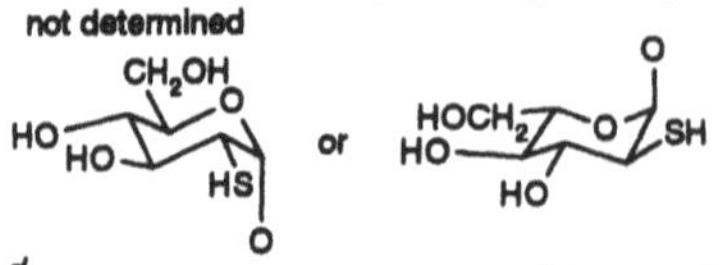

or

[d] O-D-Rhodinosyl, stereochemistry of the sugar linkage not determined

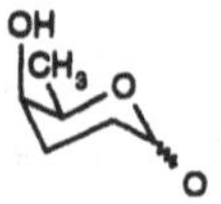

Table 6. Novel angucyclin(on)es with angular oxygens at the ring AB fusion (SF 2315 type)

	R^1	R^2	R^3	R^4
220:	H[a]	N-Ac-Cys[a,b]OH[a]	OH[a]	
221:	OH[a]	N-Ac-Cys[a,b]OH[a]	OH[a]	
224:	H	H	H	H

[a] Stereochemistry not determined

[b] N-Ac-CyS =

Table 7. Further novel angucyclin(on)es with angular oxygens at the ring AB fusion (SS-228Y and SF 2315 types)

	R^1	R^2, R^3	R^4
225:	H[a]	H[a], H[a]	H[a]
219:	OH	(double bond)	OH

[a] Stereochemistry not determined

PD 116198 (**218**), produced by *Streptomyces phaeochromogenes* WP 3688, is the enantiomer of sakyomicin B, as deduced by intensive NMR studies and a comparison of the $[\alpha]_D$ values [159]. This quite unusual fact arises from the unique biosynthesis of **218** and its several interesting congeners discussed below. One less oxygenated C-5/C-6 saturated congener **219** (here subsequently called WP 3688–2) was accumulated when *S. phaeochromogenes* was grown in the presence of the P-450 oxygenase inhibitor metyrapone [159]. Both compounds, **218** and **219**, show antibacterial activities (gram-positive/gram-negative), the former better than the latter.

Table 8. Azicemicins A and B

	R
226:	CH$_3$
227:	H

WS 009 A (**220**, = FR 901366) and B (**221**, = FR 901367), isolated from *Streptomyces* sp. No. 89009, were discovered during a screening course for endothelin receptor antagonists using ET-1, the most potent vasoconstrictor to date [160–162]. These compounds show a selective activity in an endothelin binding assay (IC$_{50}$ 5.8 × 10^{-6} M and 6.7 × 10^{-7} M, respectively). The structures feature a unique C-6a linked *N*-acetylcysteine moiety which may have arisen biosynthetically via an attack of cysteine on a 6a–12a epoxide structure, such as exemplified in A-683 (**210**) or SF 2315 B or elmycin C [3].

As the last example of the SS-228Y-type, the rhodonocardins A (**222**) and B (**223**) [163] must be mentioned here. Although already published in 1987, these interesting angucyclines were overlooked in our first review [3]. The producing organisms (*Nocardia* sp. No. 53) as well as the suggested structures are unusual because they contain a D-rhodinose (only **222**, linkage unknown), and α-glycosidically linked D-glucose and 2-deoxy-2-mercaptoglucose moieties. D-rhodinose is quite rare and was found previously only in the sakyomicins A, C and E (*Nocardia* sp. No. 53 also produces some sakyomicins); D-sugars usually occur in a β-linkage. The 4a-O and 5-O connection positions of the sugars are also novel among the angucyclines. Because of the large number of hydroxyl groups in the molecules, both wine-red colored compounds exhibit water solubility. As reported earlier for aquayamycin (**4**) and other angucyclinones, rhodonocardin B (**223**) could be acidically rearranged into a linear compound.

Four angucyclinones exhibiting only one angular hydroxy group located at AB ring fusion points were described. The structure of kanglemycin C (**224**) [164] was elucidated from spectroscopic and x-ray analysis data. No biological activity data were given. One of the major compounds within the ochracenomycin complex, ochracenomicin C (**225**, stereocenters unknown) [151], a 2,3 saturated analogue of **224**, shows weak antibacterial activities.

Structurally more interesting novel compounds within this subsection are the azicemicins A (**226**) and B (**227**). Compared with other angucyclinones, these structures are very unusual: They exhibit one angular hydroxy group (AB ring fusion) at the C-12b position, oxygens at the C-9- and C-10-positions, a keto function at C-5, a hydroquinone monomethylether (ring C), and, most importantly, an aziridine ring linked at C-3. Therefore, it has been claimed that the azicemicins represent a new structural class of antibiotics [153, 165 – 167]. Both antibiotics show moderate activities against gram-positive bacteria as well as mycobacteria (**227** better than **226**, the latter also being more toxic). The producing organism is *Amycolatopsis* sp. MJ126-NF4. The biosynthetic origin of the unusual aziridine ring, which presumably is an amino acid (serine or alanine), is still speculative. Thus the azicemicins may be the first examples of angucyclinones whose polyketide biosynthesis starts with a (modified) amino acid. As with other polyketides, amino acid starters have already been found.

One novel compound **228** (Scheme 53), here designated WP 3688 – 3, from the PD 116198 (**218**) producer *Streptomyces phaeochromogenes* WP 3688 [159], (see also above and below) exhibits a 4a,12b epoxide structure and thus is another new prototype of an angucyclinone with oxygen at the ring AB fusion. The related structure **228** shows that all hydroxy residues of this oxygen-rich compound face the same direction; **228** shows weak antibacterial activities.

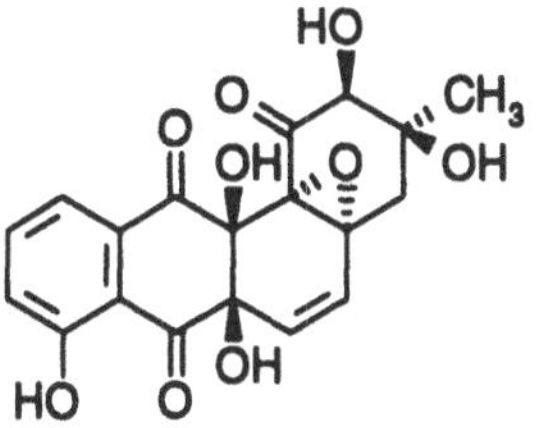

228

Scheme 53. Compound **228** (WP 3688-3), a novel angucyclinone exhibiting an unusual 4a, 12b epoxide

9.3
Typical Angucyclinones and Angucyclines Omitting Angular Oxygens at the Ring AB Fusion, but Exhibiting *C*-Glycosidic Moieties

Only some new members of this subgroup have been described (Table 9), four of them are shunt products arising from blocked mutants of the urdamycin producer *Streptomyces fradiae* Tü 2717 [164] : 104 – 1 (**229**) is the at the C-9 position *C*-glycosylated rabelomycin; 104 – 2 (**183**) is a close relative in which the OH group is "shifted" to the 5-position; 100 – 1 (**230**) is an incompletely glycosylated intermediate of the pathway leading to urdamycin B; and 124 – 1 (**231**) is 5-hydroxyurdamycin B. Their biosynthetic significance is discussed below (Sect. 10).

Balmoralmycin (**232**), isolated from *Streptomyces* sp. strain P6417, displays an inhibitory activity against protein kinase C (IC$_{50}$ 50 M), an important enzyme in context with cellular transduction mechanisms [165]. Structurally, **232** is nearly identical with the antifungal antibiotic C 104 (**197**) [153], which lacks an oxygen

Table 9. Novel urdamycin B type angucyclin(on)es

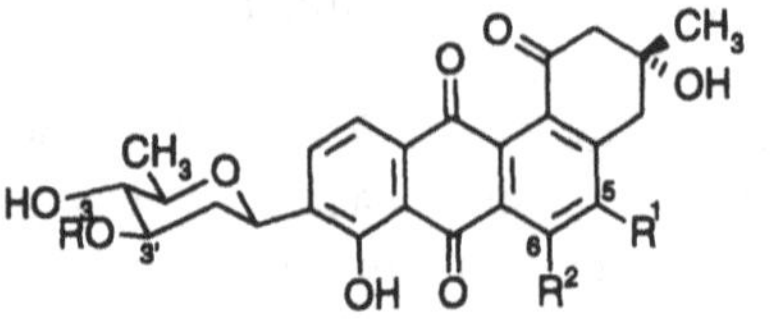

	R^1	R^2	R^3
229:	H	OH	H
183:	OH	H	H
230:	H	H	Rho^a
231:	OH	H	Oliv-Rho^b

^{a} α-L-Rhodinosyl

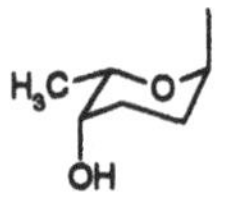

^{b} β-D-Olivosyl-1-4-α-L-Rhodinosyl

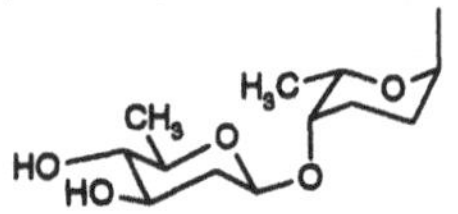

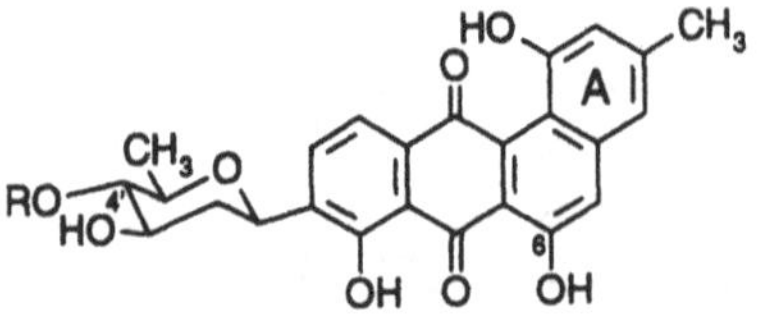

232: R =

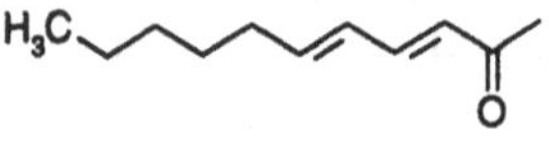

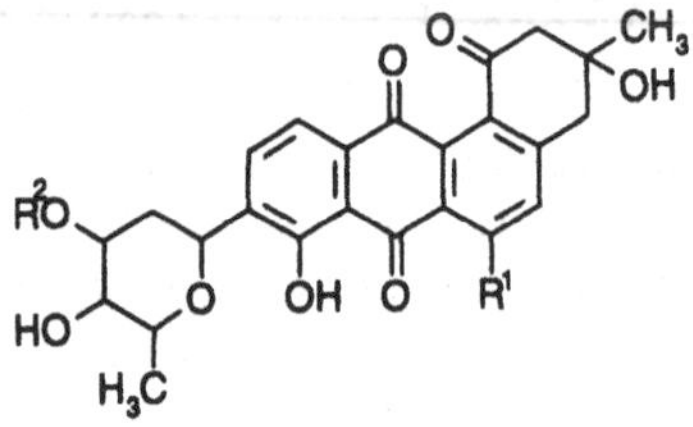

233: R^1 = H or
R^2 = ?

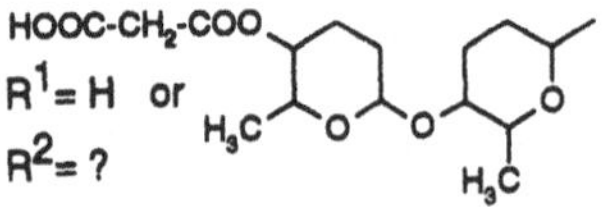

at C-6. It further resembles both the antitumor compound capoamycin, because of the (E,E)-2,4-decadienoic acid side chain at the 4′-position, and 104–1 (**229**), which is the "3,4-hydrated" analogue (without fatty acid side chain).

Some urdamycin B-type antibiotics, isolated from *Streptomyces* BA-12100, have also been patented as neoplasm inhibitors [168]. Structures **233** are (on purpose?) ambiguously drawn, and several alternatives remain (stereochemistry, R^1, R^2), which may be due to an error in the drawing of the structures.

9.4
Typical Angucyclinones and Angucyclines with Both Angular Oxygens at the Ring AB Fusion and C-Glycosidic Moieties

Most of the newly described compounds belonging to this subgroup contain aquayamycin (**4**) as the aglycon, but 2-hydroxy-, 3′-deoxy- and 4′-epi-variants were also found (Table 10).

Sch 47554 (**234**) and Sch 47555 (**235**) are two novel antifungal antibiotics from *Streptomyces* sp. SCC-2136 [167]. They were the first angucyclines containing 3′-deoxyaquayamycin (**236**) as the aglycon moiety, i.e., these compounds bear a C-glycosidically bound D-amicetose moiety instead of a D-olivose moiety in their aglyca. L-aculose and L-amicetose occur as further carbohydrate moieties.

Table 10. Novel aquaymycin type angucyclin(on)es

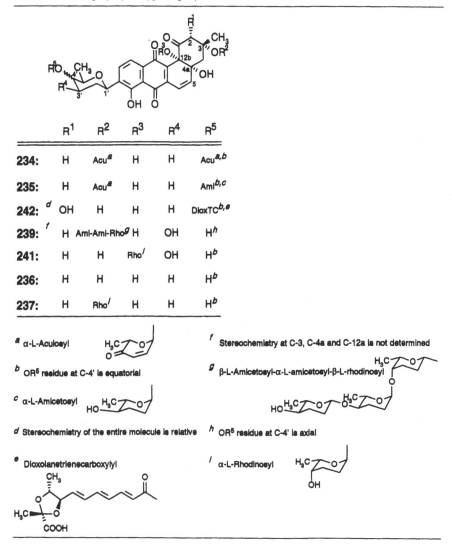

	R¹	R²	R³	R⁴	R⁵
234:	H	Acu[a]	H	H	Acu[a,b]
235:	H	Acu[a]	H	H	Ami[b,c]
242: [d]	OH	H	H	H	DioxTC[b,e]
239: [f]	H	Ami-Ami-Rho[g]	H	OH	H[h]
241:	H	H	Rho[i]	OH	H[b]
236:	H	H	H	H	H[b]
237:	H	Rho[i]	H	H	H[b]

[a] α-L-Aculosyl

[b] OR⁵ residue at C-4' is equatorial

[c] α-L-Amicetosyl

[d] Stereochemistry of the entire molecule is relative

[e] Dioxolanetrienecarboxylyl

[f] Stereochemistry at C-3, C-4a and C-12a is not determined

[g] β-L-Amicetosyl-α-L-amicetosyl-β-L-rhodinosyl

[h] OR⁵ residue at C-4' is axial

[i] α-L-Rhodinosyl

Structure **236** was also isolated from *Streptomyces* sp. strain Tü 3824 and designated as ritzamycin A, named after one of the discoverers [153]. Ritzamycin B (**237**) bears an L-rhodinose residue at C-3, and ritzamycin C is identical with Sch 47554 (**234**) [3, 169]. The ritzamycins show an interesting in vitro antitumor activity (against L1210 and HT29 cancer cells); the best results were found with **236** (A) and **234** (C). The activity is about 100- to 1000-fold better than that of aquayamycin (**4**), although **236** differs from aquayamycin (**4**) only in the 3'-residue (H instead of OH).

The novel antibiotics (active mostly against gram-positive bacteria) amicenomycins A (**239**) and B (**240**, see Sect. 9.5) were isolated from *Streptomyces* sp.

MJ 384-46 F6 [170]. The aglycon is novel, resembling aquayamycin (4) but with an inverted stereocenter at C-4' (the stereocenters at C-3, C-4a, C-12 were not determined). Also, the stereochemical connection of the deoxytrisaccharide chain is unusual, since L-rhodinose is β-linked, and L-amicetose occurs in α- as well as in β-linkage.

Compound 100-2 (241, 12b-L-rhodinosylaquayamycin), isolated from a blocked mutant of the urdamycin producer *Streptomyces fradiae* Tü 2717, represents the missing link in the glycosylation sequence of the urdamycins [164]. Structure 241 proves that the linkage of the L-rhodinose at C-12b-O is the first of three O-glycosylation steps (after the C-glycosylation of one olivose several biosynthetic steps before, see Sect. 10).

Dioxamycin (242, the shown stereochemistry is relative) [171] is a novel representative of the 4'-O-acylangucyclinones, which have a partially unsaturated C_{10} fatty acid side chain in common. But in contrast to capoamycin, balmoralmycin (232) and C104 (197), the novel acid side chain in 242 is triple unsaturated and contains an 8,9-diol structural element embedded in a dioxolane ring (most likely through a ketalization with pyruvate). The producing organism is *Streptomyces* sp. strain MH 406-SF1 which closely resembles *Streptomyces xantholiticus*. Compound 242 is active against gram-positive bacteria and various cancer cells (L1210, P388, IMC carcinoma, LX-1, SC-6), and the acute toxicity is known (LD_{50} 12.5-25 mg/kg mice, intraperitoneal injection) [171]. A presumably identical compound, designated as BE-16493 and isolated from *Streptomyces* sp. BA16493 (the published structure lacks all stereochemical information), is described in a patent as an anticancer agent [172]. Several other compounds were described, again in patent literature, thus keeping all stereochemical information open. Table 11 provides an overview.

These angucyclines show a broad variety of biological activities and possible applications: Angucyclin 243, from *Streptomyces griseolus* A-6067, is a leukemia inhibitor [173]; structure 244, from *Streptomyces* sp. CH 752 [174], was discovered during a screening for glucose-6-phosphatase inhibitors (100% inhibition at 1.2 M liter^{-1}, IC_{50} 0.6 M liter^{-1}). This enzyme plays a major role in carbohydrate metabolism in the liver. Angucycline 245, from *Streptomyces griseovirides* A-7884, is described [175] as a useful low density lipoprotein (LDL) uptake promoter (3.0 g ml^{-1} increases LDL uptake into human hepatoma culture cells by 237%). Some members of the BA-12100 complex of antibiotics [168] with antineoplastic activities were already mentioned in Sect. 9.3: structures 246 and 247 are aquayamycin (4)-type variants within this complex. Another large complex of aquayamycin (4)-type angucyclines is 248 (structures not further specified); these were isolated from *Streptomyces nodosus* (FERM P-7985) and claimed as antitumor agents active against multidrug-resistant cells [176].

9.5
Natural Products Biosynthetically Derived from Angucyclinones and Angucyclines with Rearranged Skeletons Initiated by Oxidative Biosynthetic Processes

Looking only at the chemical structures, the natural products described in this subsection would not belong to the angucycline group. However, these natural

Table 11. Novel aquaymycin type angucyclin(on)es with unspecified chemistry

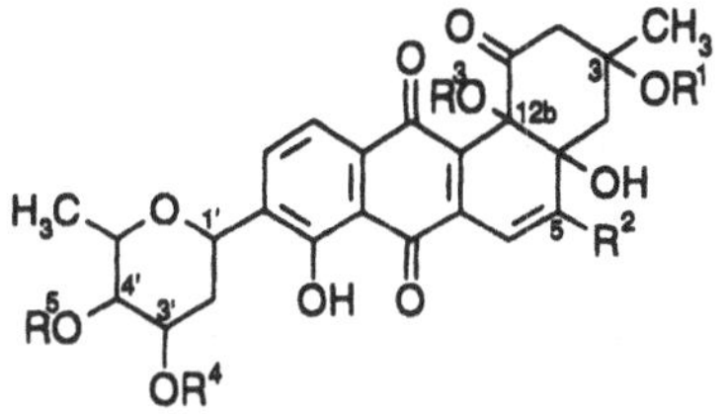

	R[1]	R[2]	R[3]	R[4]	R[5]
243:	H	NAcCys[a]	H	H	DS[b]
244:	DS[c]	H	MS[d]	H	DS[c]
245:	H	H	H	H	DS[c]
246:	H	H	H, MS[e]	ADS[f]	H
247:	H	H	H, MS[e]	DS[g]	H
248a:	H, DS[h], DS[i]	H	H	H	MS[d], MS[j]
248b:	H, DS[h], DS[i]	H	H	MS[k]	

[a] N-Acetylcysteinyl:

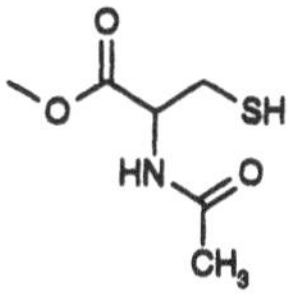

[b] Disaccharyl residue:

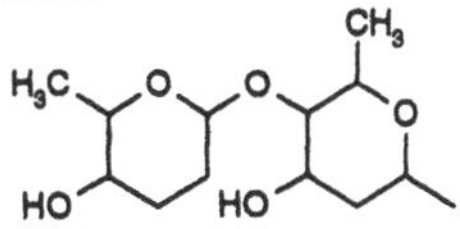

[c] Disaccharyl residue:

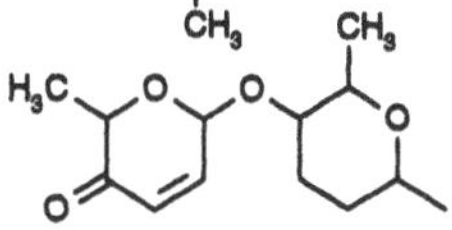

[d] Monosaccharyl residue:

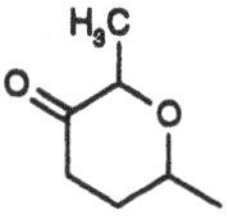

[e] Monosaccharyl residue:

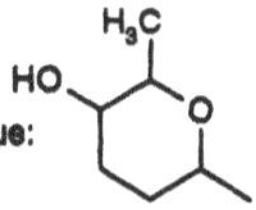

[f] Acyldisaccharyl residue:

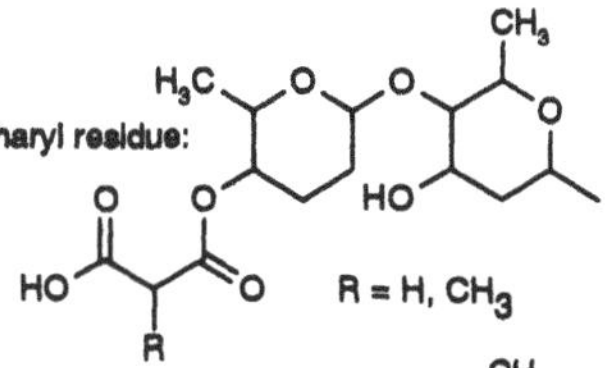

R = H, CH₃

R = H, CH₃ → $R = H, CH_3$

[g] Disaccharyl residue:

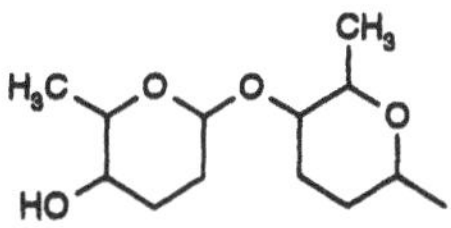

[h] Disaccharyl residue:

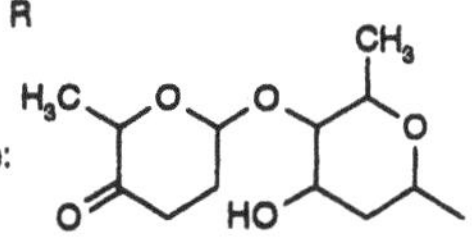

[i] Disaccharyl residue:

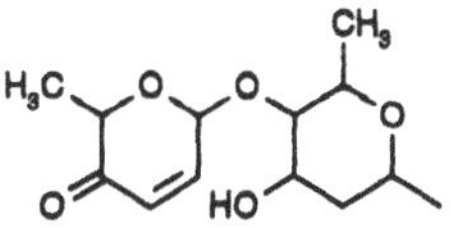

[j] Monosaccharyl residue:

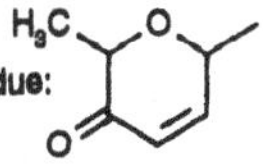

[k] Monosaccharyl residue:

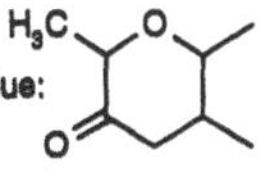

products were proven to derive biosynthetically from typical angucyclinones or angucyclines and are thus included in this review.

Again, "derivatives" of some of the above mentioned aquayamycin (**4**)-type angucyclines exhibiting an opened ring A were found (Table 12). Their relationship to their above mentioned congeners is analogous to that of vineomycin A_1 to B_2 [3, 169].

Ritzamycin D (**238**) [153] and amicenomycin B (**240**) [170] are ring A opened variants of ritzamycin C (**234**, Sch 47554) and amicenomycin A (**239**), respectively. These congeners probably arise biosynthetically from an oxidative C-1/C-12b cleavage step of the ring A closed analogues late in biosynthesis. Like vineomycin B_2 [169] and in contrast to ritzamycin D (**238**), **240** is the free acid, not the methyl ester.

Two more (see also above) new natural products related to PD 116198 (**218**) were isolated from *Streptomyces phaeochromogenes* WP 3688 [159], these are referred to here as WP 3688–4 (**249**) and WP 3688–5 (**250**) (Scheme 54), the chemical formulae representing their relative stereochemistry. These compounds display novel molecular backbones containing unusual oxygen heterocycles. The producing strain possesses an unusual repertoire of (obviously not too specific) oxygenases which cause a mechanistically not yet understood formation of these novel oxidized skeletons. Interestingly, these compounds were found during investigations of *Streptomyces phaeochromogenes*, using mono-

Table 12. Novel ring A opened (vineomycin B type) angucyclines

	R^1	R^2	R^3	R^4	R^5
240:	H	Ami-Ami-Rho[a,b] OH	OH	H	OH
238:	CH_3	O-Acu[c]	H	O-Acu[c]	H

[a] β-L-Amicetosyl-α-L-amicetosyl-β-L-rhodinosyl

[b] Stereochemistry at C-12 is not determined

[c] α-L-Aculosyl

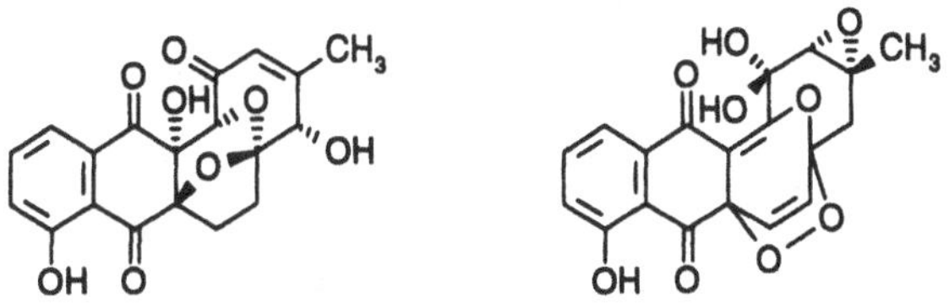

Scheme 54. Two novel, through oxygenase(s) modified, angucyclinones: 249 (WP 3688-4) and 250 (WP 3688-5)

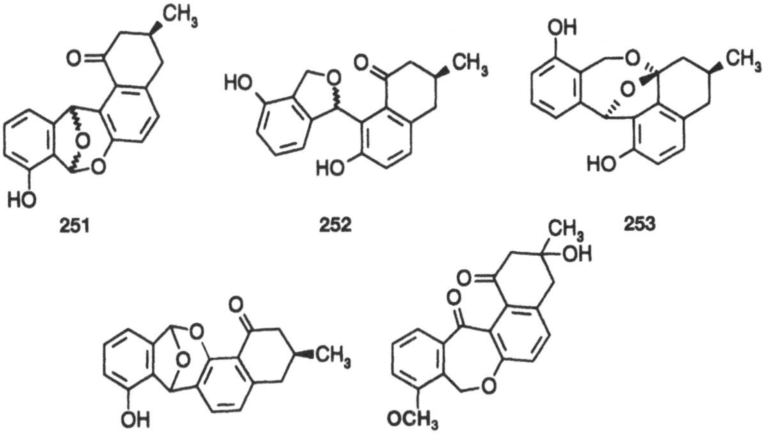

Scheme 55. Novel, through oxygenases modified, angucyclinones: emycins D–G (251–254)

and dioxygenase inhibitors, which were intended to complement biosynthetic studies on the origin of the oxygens of PD 116198 (218) and to explain the biosynthetic formation of its oxygen-rich ring A.

Similar, interesting, novel oxygen heterocyclic structures resulted from a mutation of the emycin producer *Streptomyces cellulosae* ssp. *griseoincarnatus* (strain FH-S 1114) [155–157]. The normal products of the wild-type strain were ordinary angucyclinones, such as ochromycinone (35), 3-deoxyrabelomycin (20), emycin A and elmycin D [3]. The mutant 1114–2 had a widely changed product spectrum including the novel emycins D (251), E (252), F (253), and G (254) (Scheme 55). Biosynthetic studies (see below) on the major products of *Streptomyces cellulosae* ssp. *griseoincarnatus* (mutant 1114–2) showed that presumably a modified oxygenase activity and subsequent nonenzymatic reactions are responsible for the formation of the novel products and explain also the occurrence of diastereomeric mixtures for 251 and 252.

Angucyclinone C (255) from *Streptomyces* sp. strain Gö 40/14 [153] exhibits a novel seven-membered ether ring and is (like the emycins D 251, E 252 and F 253) a ring C-modified angucyclinone. Compound 255 is speculated to be a consecutive product generated by an initial attack of 7-OH on the epoxide carbon C-6a. Thus the formation of **255** is also caused initially by a biosynthetic oxidation reaction.

The structures of the kinamycins were revised recently [177, 178], as they were shown to be 5-diazobenzo[*b*]fluorenes, e.g. **256** (kinamycin D), and not cyanocarbazoles as originally designated. The biosynthetic relation of the kinamycin antibiotics to the angucycline group was established by biosynthetic studies revealing dehydrorabelomycin (**257**) as their biosynthetic intermediate (Scheme 56) [3, 179].

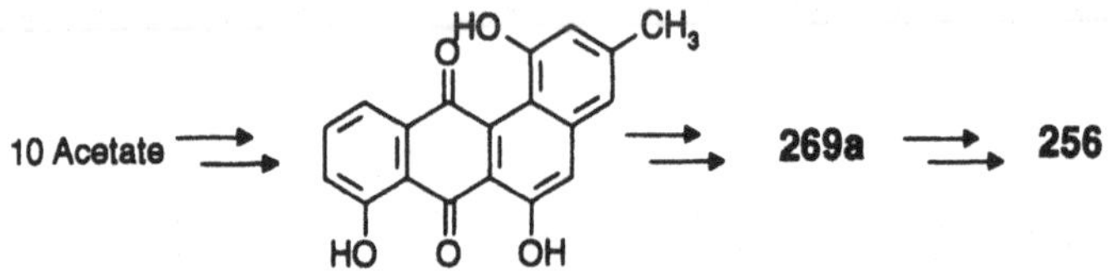

Scheme 56. Dehydroabelomycin (**257**), a biosynthetic angucyclinone intermediate of the kinamycins

Table 13. Novel kinamycin benzo[*b*]fluorene antibiotics, biosynthesized via the angucyclinone dehydrorabelomycin (**257**)

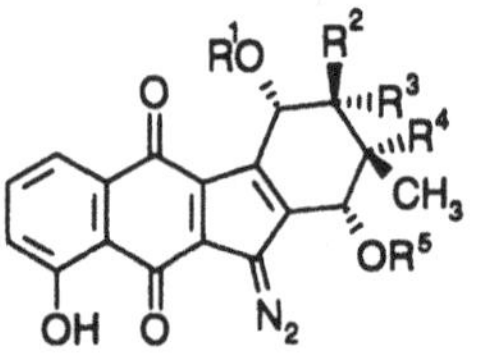

	R^1	R^2	R^3	R^4	R^5
256:	H	$OCOCH_3$	H	OH	$COCH_3$
258:	$COCH_3$	$OCOCH(CH_3)_2$	H	OH	$COCH(CH_3)_2$
259:	$COCH_3$	$OCOCH_3$	H	OH	$COCH(CH_3)_2$
260:	$COCH_3$	$OCOCH_3$	H	$OCOCH(CH_3)_2$	$COCH(CH_3)_2$
261:	H	$OCOCH_3$	H	$OCOCH(CH_3)_2$	$COCH_3$
262:	H	H	O (Oxirane)		$COCH(CH_3)_2$
263:	H	H	O (Oxirane)		$COCH_3$
264:	H	$OCOCH_3$	H	OH	$COCH(CH_3)_2$
265:	H	$OCOCH_3$	H	OH	$COCH_2CH_3$
266:	H	OH	H	OH	$COCH(CH_3)_2$

Several new kinamycin-related natural products have been discovered [180–182] since our first review [3]. Some of the structures are summarized in Table 13; all are corrected here to diazofluorene structures (A83016A **258**, 4-deacetyl-4-O-isobutyrylkinamycin C **259**, 3-O-isobutyrylkina-mycin C **260**, FL-120A **261**, FL-120B **262**, FL-120B′ **263**, FL-120C **264**, FL-120C′ **265**, FL-120D **256** = kinamycin D, FL-120D′ **266**).

Since the kinamycins turned out to be benzofluorene derivatives, several other benzofluorene derivatives have to be considered to be biosynthetically related to the kinamycins and thus to the angucycline group of natural products, too (see also Sect. 10). Some recent examples are depicted in Scheme 57.

Kinafluorenone (**269**) and kinobscurinone (**269a**) were discovered as a major product of a mutant of the kinamycin producer *Streptomyces murayamaensis* which is blocked in kinamycin biosynthesis [183]. The latter compound was shown to be a more advanced intermediate of kinamycin biosynthesis [183]. Cysfluoretin (**270**) [184], the stealthins A (**267**, CA 39 A), B (**268**, CA 39 B) [185, 186], the momofulvenones A (**271**), B (**272**) [187] and Tü 96 OR (**273**) [188] (Table 14) are metabolites of various other *Streptomyces* sp. which do not produce kinamycins. Incorporation experiments with ^{13}C-labeled acetate give evidence of a relation of the momofulvenones to the kinamycins [187]. Compound **273** is a product of a blocked mutant of the lysolipin-producer *Streptomyces violaceo-niger* Tü 96. There is no obvious biosynthetic relation between the dodecaketide lysolipin I (**274**) [189] and the decaketide Tü 96 OR (**273**), unless one assumes a random effect on the chain length determining factor [190–194] of polyketide biosynthesis through the nitrosoguanidine mutagenesis experiment (for alternative possibilities see Sect. 10).

Finally, the jadomycins A (**275**) and B (**276**), novel benz[*b*]oxazolophenanthridine antibiotics from *Streptomyces venezuelae* ISP 5230 [195, 196], have to be included here (Scheme 58). The glycosylated jadomycin B (**276**) was produced by *Streptomyces venezuelae* ISP 5230 under stress conditions (heat shock or ethanol treatment) [196]. Based on the results of biosynthetic studies on the

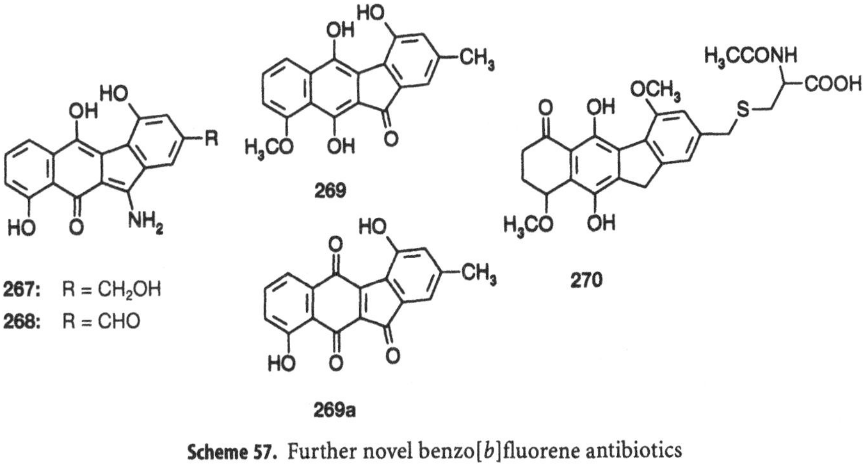

267: R = CH₂OH
268: R = CHO

Scheme 57. Further novel benzo[*b*]fluorene antibiotics

Table 14. Momofulvenones A,B (272,271) and Tü96OR (273), a natural product from a blocked mutant of the lysolipin I (274) producer *Streptomyces violaceoniger* Tü9

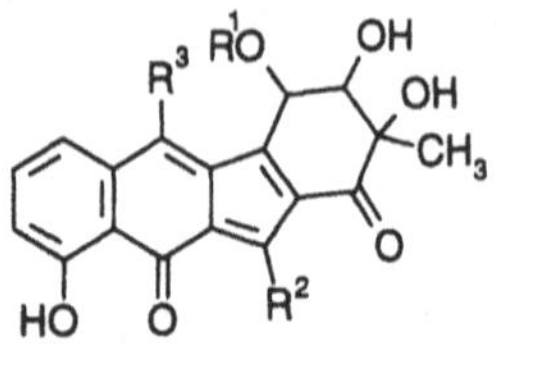

	R^1	R^2	R^3
271:	H	H	OM^{+a}
272:	COCH$_3$	H	OM^{+a}
273:	H	OH	H

^{a} Various Cations

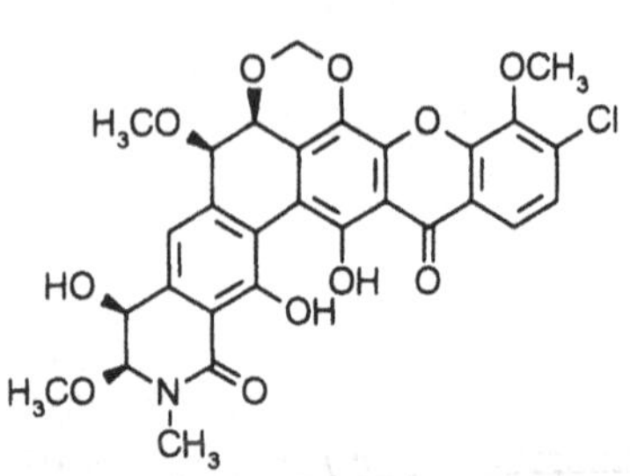

274

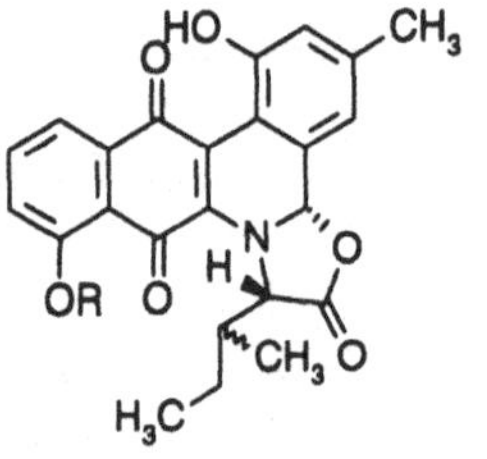

275: R = H

276: R =

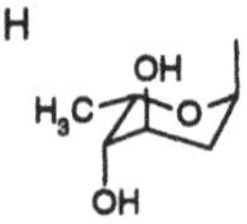

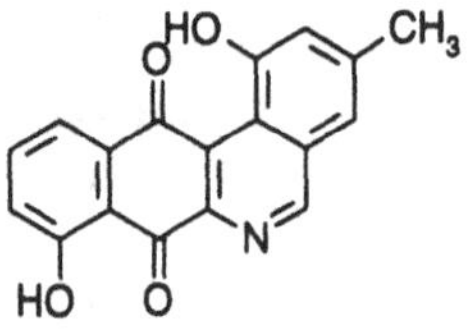

277

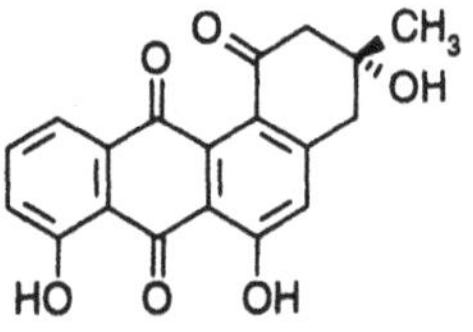

29

Scheme 58. Jadomycins A,B (275, 276) and their biosynthetic intermediate rabelomycin (29)

kinamycin antibiotics and especially due to the analysis of their biosynthetic gene cluster [197] (see also Sect. 10), **275** and **276** are associated biosynthetically with the angucycline group compounds: The phenanthroviridin aglycon (**277**), a compound closely related to a proposed intermediate of the kinamycin biosynthetic pathway [198], was detected in the fermentation broth of the same strain [195]. Later, **277** was found as a minor product of an UV mutant of the kinamycin producer *S. murayamaensis* [199]. A disruption of an oxygenase gene of the jadomycin biosynthesis resulted in the production of rabelomycin (**29**) [200]. The latter experiment proves unambiguously the biosynthetic relationship of the jadomycin group to the angucycline group (for further details see Sect. 10).

10
Biosyntheses

In this section, recent (1991 to mid-1996) biosynthetic studies related to the angucycline group of antibiotics will be discussed. First, the biosynthetic formation of the tetracyclic angucyclinone frame will be emphasized. This will be followed by studies of biosynthetic gene clusters of angucycli(no)nes and on late biosynthetic steps of angucycline group antibiotics, i.e. those steps caused by post-polyketide modifying enzymes.

10.1
Studies on the Formation of the Tetracyclic Angucyclinone Frame

In general, the tetracyclic benz[a]anthracene backbone of the angucyclines and angucyclinones derives from ten "acetate" (i.e., one acetyl CoA starter and nine malonyl CoA extender) units which are condensed in the typical head-to-tail fashion of the polyketide biosynthetic pathway [3, 201, 202]. As the first exception, the azicemicins [153, 165–167] **226** and **227** can be considered; these most likely derive from one (modified) amino acid and nine acetates (see above, Sect. 9.2). However, no studies have yet been published on the biosynthesis of the azicemicins. In contrast to the biosynthetically related anthracyclines, polyketide biosyntheses of the angucyclinones studied so far are always initiated with acetyl CoA, and not with propionyl CoA or other CoA-activated fatty acids. However (see Sect. 10.3) some members of the gilvocarcin group, the just recently published (after this article was finished) brasiliquinones A–C [203] and a synthetically prepared angucycline/anthracine hybrid [37] exhibit propionate starter units. In the general "standard" pathway, the ten acetate units are incorporated into the benz[a]anthracene backbone starting with C-13 and ending with C-2 under decarboxylation of the last acetate (Scheme 59, e.g., for the formation of elmycin D **278**).

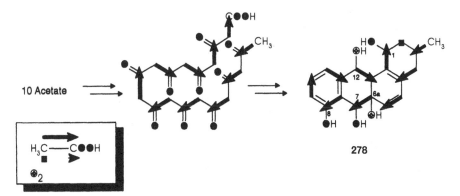

Scheme 59. Typical biosynthetic formation of an angucyclinone, e.g., elmycin D (**278**), from acetate and molecular oxygen

The biosynthetic pattern shown in Scheme 59 for compound **278** was first found for vineomycin A_1, thereafter for several other angucyclin(on)es [3], and more recently for dehydrorabelomycin (**257**), elmycin D (**278**), emycin A, landomycin A (**208**), ochromycinone (**35**), and PD 116740 (**279**, see Scheme 60). As an element of novelty, studies on the biogenetic origin of the oxygen atoms were included in these studies by feeding experiments with $[1-^{13}C^{18}O_2]$acetate and growth of the cultures under an $^{18}O_2$-enriched atmosphere. While the studies on aquayamycin (**4**), dehydrorabelomycin (**257**), emycin A, elmycin D (**278**), and ochromycinone (**35**) showed the expected results [155, 156, 204, 205], i.e., all oxygens attached to a former C-1 of an acetate unit deriving from acetate, all other oxygens (attached to a former C-2 of an acetate unit) originate from molecular oxygen (see Scheme 59 for **278**). However, some of the oxygens are exchangeable (see below). Exceptions were found for landomycin A (**208**) [147] and PD 116740 (**279**) [54, 206] indicating unexpected, prearomatic deoxygenation steps (see Scheme 60).

The oxygen linked at the 6-position of PD 116740 (**279**) is assumed to derive from water, since an epoxidation and subsequent enzymatic hydrolysis is assumed for the biosynthetic sequence leading to **279** [54, 206]. The biogenetic origin of the oxygen at the 6-position in landomycin A (**208**) could not be proven to stem directly from aerial oxygen, since the fermentation of *Streptomyces cyanogenus* S-136 under ^{18}O atmosphere yielded only 5,6-anhydrolandomycin A [147]. But since this oxygen was not labeled by $[1-^{13}C,^{18}O_2]$ acetate, and 5,6-anhydrolandomycin A turned out to be the major product in a fermentation

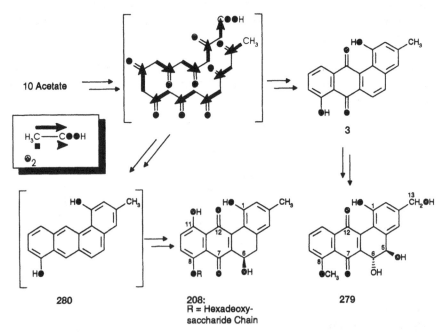

Scheme 60. The unexpected biogenetic origin of oxygen atoms in landomycin A (**208**) and PD 116740 (**279**) indicates prearomatic deoxygenation steps

under reduced oxygen [204, 207], it is very likely that 6-O of **208** derives from molecular oxygen.

It remains obscure why these deoxygenation steps occur during biosynthesis, i.e., why nature sometimes leaves the usual straightforward biosynthetic approach and seems to waste energy by removing oxygens from certain positions which in later biosynthetic steps have to be reintroduced through oxygenases. It is also not completely clear whether these deoxygenations happen during the assembly of the polyketide chain (as established for the type I polyketides, consistent with their modular gene cassettes containing ketoreductases and dehydratases) or (directly) after the formation of the polyketide chain, i.e., through ketoreductases and aromatases (as established for typical type II polyketides). The former would place those natural products in which prearomatic deoxygenations were observed closer to the type I polyketides. Since angucyclines are structurally typical type II polyketides (polycyclic aromatic polyketides), the latter possibility is favored. Also, analysis of the biosynthetic genes of landomycin A (**208**) by Bechthold et al. [208] yielded a typical type II polyketide synthase (PKS) cluster (see below), and not the modularly assembled PKS genes which are typical for type I polyketides [208–211]. Finally, also postaromatic deoxygenations cannot be ruled out completely and were already demonstrated on plant type II polyketides [212]. For the biosynthesis of landomycin A (**208**) and PD 116740 (**279**), aromatic biosynthetic intermediates **280** (for **208**) and **3** (for **279**) were postulated and determined, respectively [54, 147, 206]. The analysis of oxygen biogenesis with ^{18}O-labeled precursors ($CH_3C^{18}O_2H$, $^{18}O_2$; an incorporation is indicated by upfield ^{18}O-induced shifts of ^{13}C resonances of directly ^{18}O-linked carbons) is further complicated because of a possible oxygen exchange at carbonyl groups, such as C-7 and C-12. The ^{18}O label may be washed out in the aqueous fermentation medium via the hydrate. For instance, expected labels in one or the other feeding experiment were decreased in the 7-position of dehydrorabelomycin (**257**) and could neither be detected at the 12-position in urdamycinone A (**4**, aquayamycin), emycin A and in ochromycinone (**35**), nor in the 7- or 12-positions of PD 116740 (**279**) and PD 116198 (**218**, see below) [54, 96, 157, 159, 204, 206, 213, 214].

A completely novel biosynthetic assembly of an angucyclinone and evidence for a rearranged molecular skeleton were discovered during biosynthetic studies of angucyclinone PD 116198 (**218**). Incorporation experiments using singly and doubly ^{13}C-labeled acetate resulted in a pattern that is consistent with a biosynthesis leading initially to an anthracyclinone, which then rearranges into the angucyclinone under the influence of post-polyketide modifying enzymes [96, 214]. An alternative, namely, the possibility of a two-chain biosynthesis, could be excluded because of negative results from feeding experiments with labeled orsellinates. Regarding the oxygen biogenesis, it is most surprising that 3-OH derives from molecular oxygen, and not from acetate. Scheme 61 shows the results of all feeding experiments with ^{13}C- and ^{18}O-labelled precursors and the concluded biosynthetic hypothesis. Additional incorporation experiments with deutero-labelled precursors were carried out which also contributed to the conclusion of the rearrangement hypothesis shown in Scheme 61. Several novel metabolites arose from further studies with P-450 oxygenase inhibitors [159],

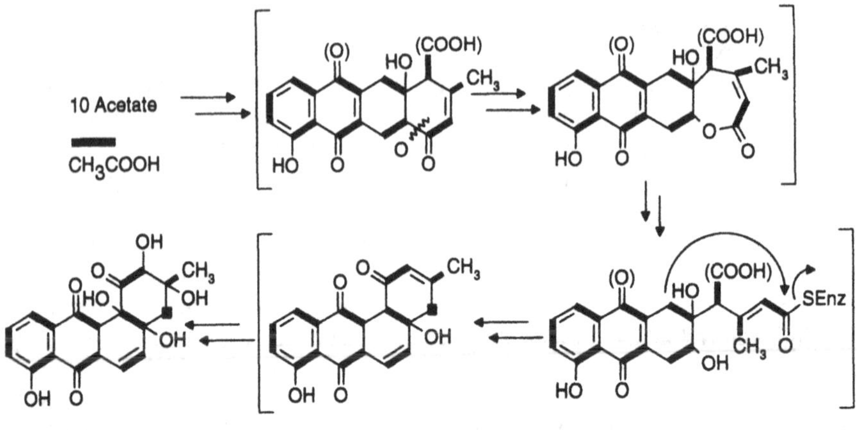

Scheme 61. Biosynthetic studies on PD116198 (**218**) reveal an unusual rearrangement of an anthracyclinone into an angucyclinone, caused by an oxygenase

one of them (**219**) can be considered a shunt product deriving from an immediate biosynthetic precursor of **218** (see above), although no linear decaketides, such as anthracyclinones, could be detected.

Another new variant of the biosynthesis of the angucyclinone skeleton may be BE-23254 (**200**) [141] (see Sect. 9.1), since its biosynthesis may start with formyl CoA or malonyl CoA (in the latter case using only nine ketide building blocks) and it does not undergo decarboxylation.

10.2
Gene Clusters of the Biosynthesis of Angucyclin(on)es and
Studies of Biosynthetic Steps Caused by Post-Polyketide Modifying Enzymes

Modern biosynthetic studies cannot be accomplished without considering the DNA level, i. e., studying the genes coding for the biosynthetic enzymes. Several type 2 polyketide producing organisms have already been analyzed with respect to the biosynthetic gene clusters of their main secondary metabolites. Among these are three angucycline producers: *Streptomyces venezuelae*, the producer of jadomycin B (**276**) [197, 200]; *Streptomyces fradiae*, the urdamycin A (**281**) producer [215–217]; and *Streptomyces cyanogenus*, the producer of landomycin A (**208**) [208]. All identified biosynthetic genes are representatives of typical type II PKSs, which are multifunctional enzyme complexes consisting of subunits that are genetically coded by various iteratively used open reading frames (ORFs). The key elements are the ketosynthase/acyltransferase (KS/AT), the chain length factor (CLF) and the acyl carrier protein (ACP), altogether the so-called minimal PKS. Typical additional elements of the PKS are ketoreductases (KRs), cyclases (CYCs) and aromatases catalyzing successively ketoreductions and dehydration reactions (AROs). Most important among the genes coding for post-polyketide modifying enzymes are oxygenases (OXs) and group transferases, e.g., methyl transferases (MTs) or glycosyl transferases (GTs). The genes so far identified are shown in Scheme 62. In *Streptomyces cyanogenus and Streptomyces fradiae*, additional genes of the deoxy sugar biosyntheses and of the glycosyl transferases were also recognized [208, 218].

Thus from an analysis of the biosynthetic gene clusters as well as from earlier biosynthetic studies (often with blocked mutants), it is useful to divide the polyketide biosynthetic pathway into PKS reactions and post-PKS tailoring steps. Several of the latter were investigated with respect to biosyntheses of angucyclin(on)es and natural products derived thereof. Intensive studies of various blocked mutant products of *Streptomyces fradiae* Tü 2717 resulted in a well established sequence of the late steps of urdamycin A (**281**) biosynthesis (Scheme 63) [164].

It is surprising that one glycosyl transfer step (establishing the *C*-glycosidic moiety) occurs prior to several other modification steps, in particular an oxygenation, a ketoreduction and a dehydration (no gene has yet been identified responsible for this step), before final glycosyl transfers complete the molecule **281**. Urdamycin A (**281**) is then further modified with amino acids into the dark, discolored urdamycins C, D, E and H [3].

Landomycin A (**208**) consists of the aglycon landomycinone and an unusual phenolglycosidically linked hexadeoxysaccharide chain. The latter twice combines the sequence oliv-4-1-oliv-3-1-rho (in which oliv is D-olivose, and rho L-rhodinose). Thus for the biosynthesis of the hexadeoxysaccharide chain (at least) two alternatives were possible: (1) initial biosythesis of the trisaccharide oliv-oliv-rho which is then subsequently attached to landomycinone and the landomycinone trisaccharide intermediate, respectively, and (2) the successive linkage of one sugar moiety after the other, a sequence of glycosylation steps which is completed with the formation of landomycin A (**208**). In the first alternative, landomycin A would be biosynthesized hither to landomycin B (**282**) and D (**283**), which subsequently would be formed through the influence of glycosylases. In the second alternative, **282** and **283** should be intermediates of the biosynthesis of **208** (Schemes 64, 65).

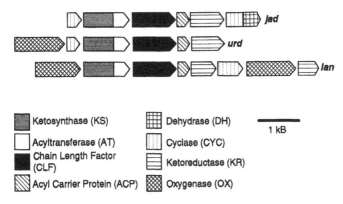

Scheme 62. Genes coding for the polyketide synthases of angucyclinones

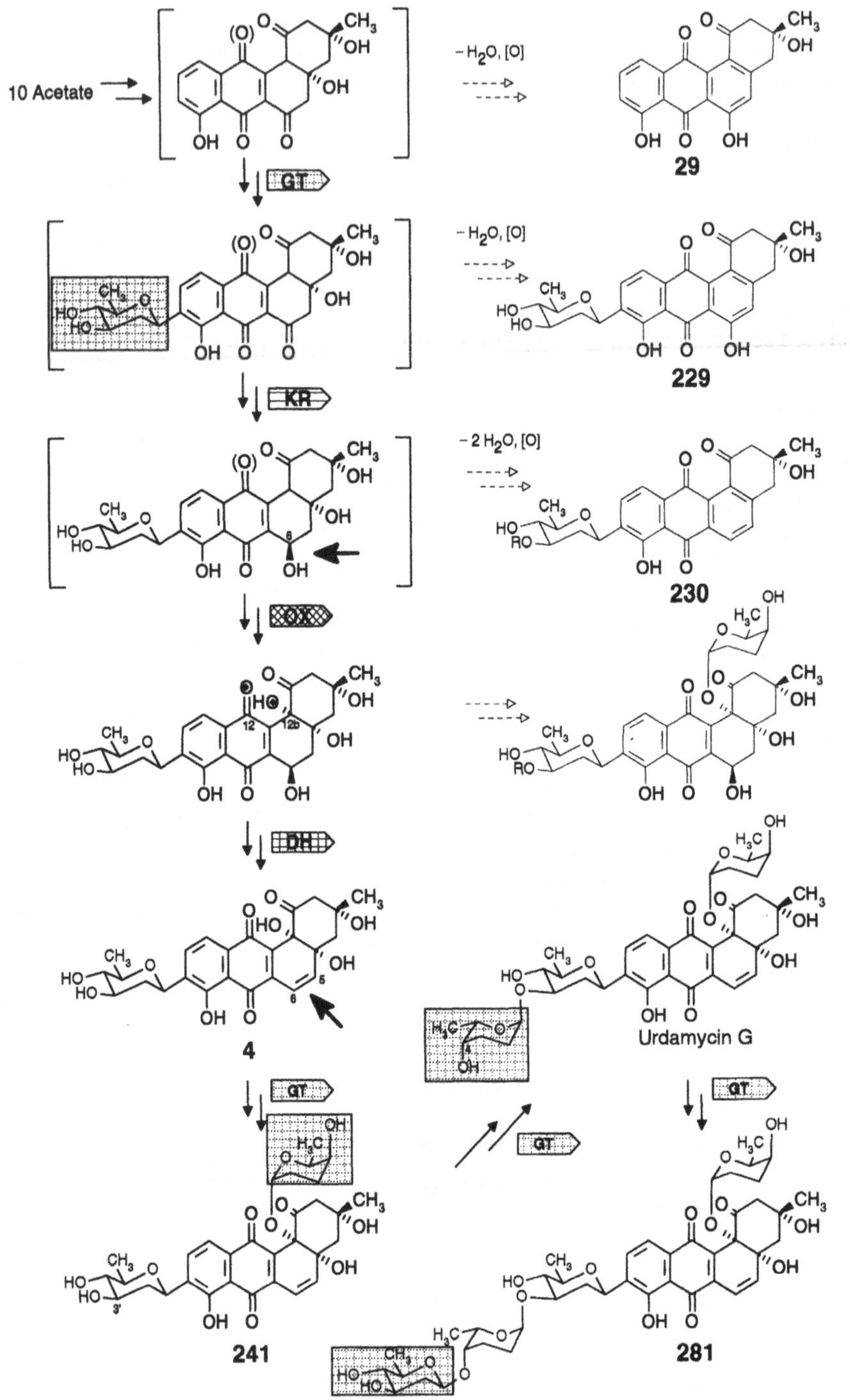

Scheme 63. Late biosynthetic steps of the urdamycin A (**281**); this scheme is supported by the isolation of the shunt products **29**, **229** and **230**, as well as the intermediates **4**, **241** and urdamycin G (unnumbered)

Landomycinone
OH O HO CH3
O O OH
Olivose
Olivose
Rhodinose
Olivose
Olivose
Rhodinose
A
B
C
D
E
F
208
208: Landomycinone, Sugars A, B, C, D, E, F
282: Landomycinone, Sugars A, B, C, D, E
283: Landomycinone, Sugars A, B
284: Landomycinone, Sugars A, B, C

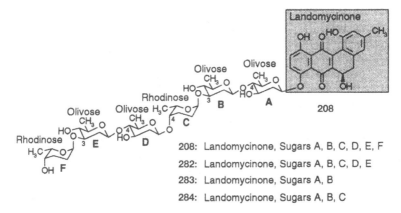

Scheme 64. The landomycins A, B, D and E (**208, 282–284**) differ only with respect to the length of their glycan chain

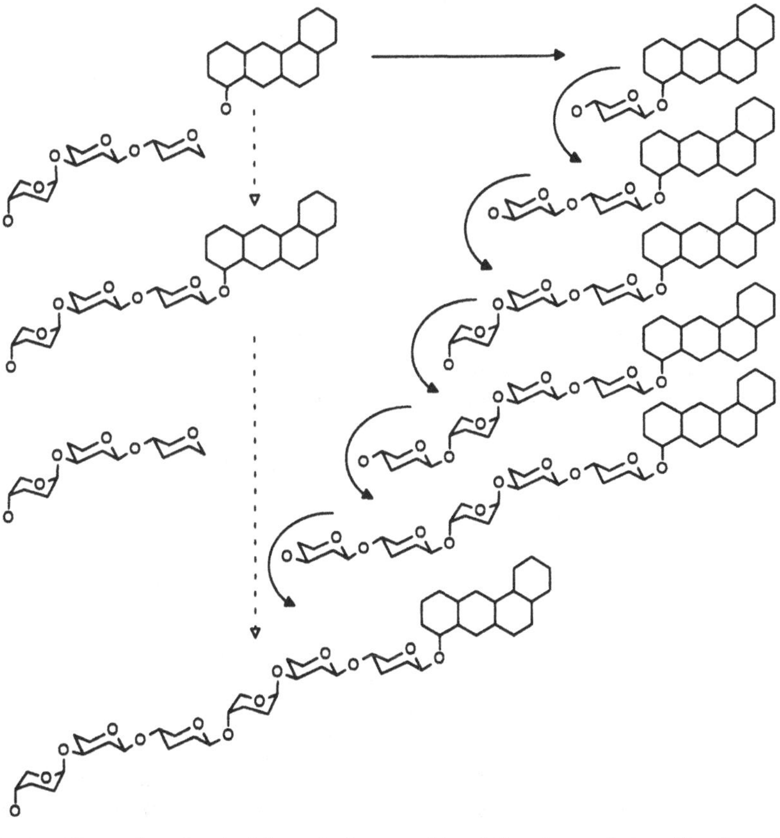

Scheme 65. Alternatives for the biosynthetic assembly of the glycan chain of landomycin A (**208**); the alternative depicted on the right side (*solid arrows*) is correct

Cross-feeding experiments with various landomycins, all ^{14}C-labeled via their biosynthesis using [1-^{14}C]acetate, confirmed the second alternative, because incorporations of landomycin D (283) into the landomycins B (282) and A (208) as well as of 282 into 208 could clearly be observed; by contrast, radiolabeled 208 was not incorporated into any of the other landomycins [219–221]. Interestingly, the landomycinone monosaccharide and the trisaccharide 284 could not be detected as metabolites of the landomycin producer *Streptomyces cyanogenus*. The latter (284), however, subsequently called landomycin E, is the major product of *Streptomyces globisporus* 1912, a strain which was genetically manipulated with the putative antibiotic production-stimulating plasmid pSG 1912 [222–224]. For the biosynthetic sequence leading to landomycin A in *S. cyanogenus* it is assumed that either the disaccharide oliv-oliv is formed prior to its glycosylation or, more likely, that the glycosylation steps which transfer a D-olivose moiety are much faster than the glycosyltransfers with L-rhodinose. Interesting oxygenases are important biosynthetic enzymes for the formation of the kinamycins [3, 177, 178, 183, 199, 225] and the jadomycins [195, 196, 197, 200]. Both biosyntheses involve an angucyclinone intermediate (257) that undergoes, in a key step, an oxidative ring B fission (presumably into 285) before decarboxylation, ring closure and a condensation reaction with molecular nitrogen (or hydrazine, or transamination and subsequent reaction with NO?) and the amino acid isoleucine, leading to kinamycin F (286, via kinobscurinone 269a and prekinamycin 287) and jadomycin B (276), respectively (a possible sequence is shown in Scheme 66).

A mutation, caused by the influence of the P-450 oxygenase inhibitor metyrapone, of one (or more) of the oxygenases which are normally involved in PD 116198 (218) biosynthesis may be the reason for the formation of the recently discovered novel natural products WP 3688–3 (228), WP 3688–4 (249), and WP 3688–5 (250) mentioned in Sects. 9.2 and 9.5, respectively, although no rational explanation for their formation has been published yet [159]. For the generation of the novel blocked mutant products emycins D–G (251–254, see Sect. 9.5), mutation of a key oxygenase of *Streptomyces cellulosae* ssp. *griseoincarnatus* seems to be likely. This mutation causes either a biological Baeyer-Villiger oxidation of ochromycinone (35) [155, 156] or a 1,3 diol cleavage of a postulated emycin H (216) analogue 288; both mechanisms are currently being examined (Schemes 67, 68) [156, 157].

10.3
Conclusions

Oxidoreductases seem to be quite substrate-flexible enzymes in the biosyntheses of angucyclinones, as is clear from biosynthetic studies on PD 116198 (218), the WP 3688 compounds [96, 159, 214] and the emycins [155, 156]. Oxygenases in particular cause more dramatic effects besides the obvious introduction of oxygen atoms. These include: C-C bond cleavages, e.g., via a biological Baeyer-Villiger oxidation, as discussed for the biosyntheses of PD 116198 (218) and emycin F (253); rearrangements; and the introduction or change of stereochemical features. In the biosynthetic formations of the kinamycins and of the

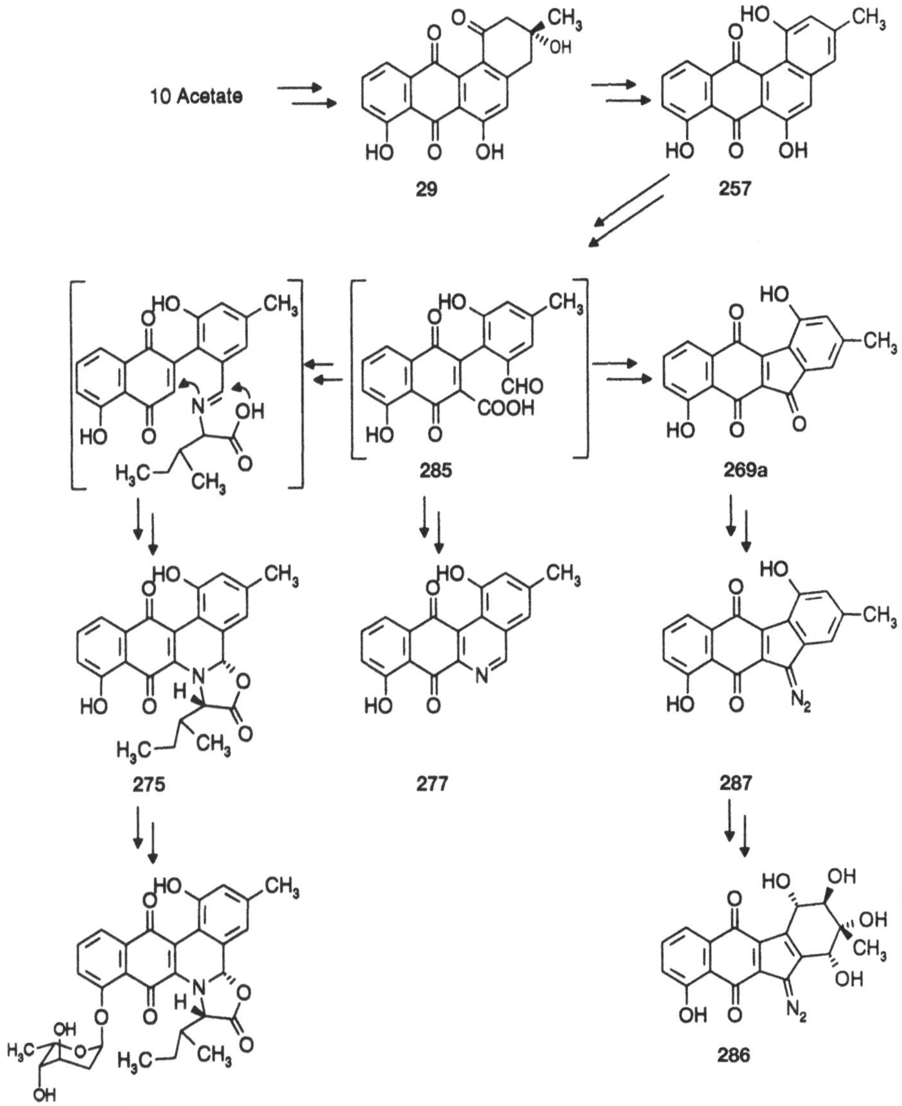

Scheme 66. Hypothetical biosynthetic pathway leading to the phenanthroviridin aglycon (277), the jadomycins (275, 276) and the kinamycin family, e.g., kinamycin F (286). The biosynthetic key step is the oxidative 5,6-bond fission, presumably from dehydrorabelomycin (275) to 285

jadomycins, oxygenase-induced C-C-bond cleavage is a key biosynthetic step. Although it is possible to convert aquayamycin (4) chemically into vineomycinone B by treatment with acids [3], the latter also may arise biosynthetically through an oxidative C-C bond cleavage. Also the very interesting antitumor antibiotics of the gilvocarcin class [226–241], e.g., the gilvocarcins M (289), E (290), V (291; toromycin, anandimycin) and defucosylgilvocarvin V (292),

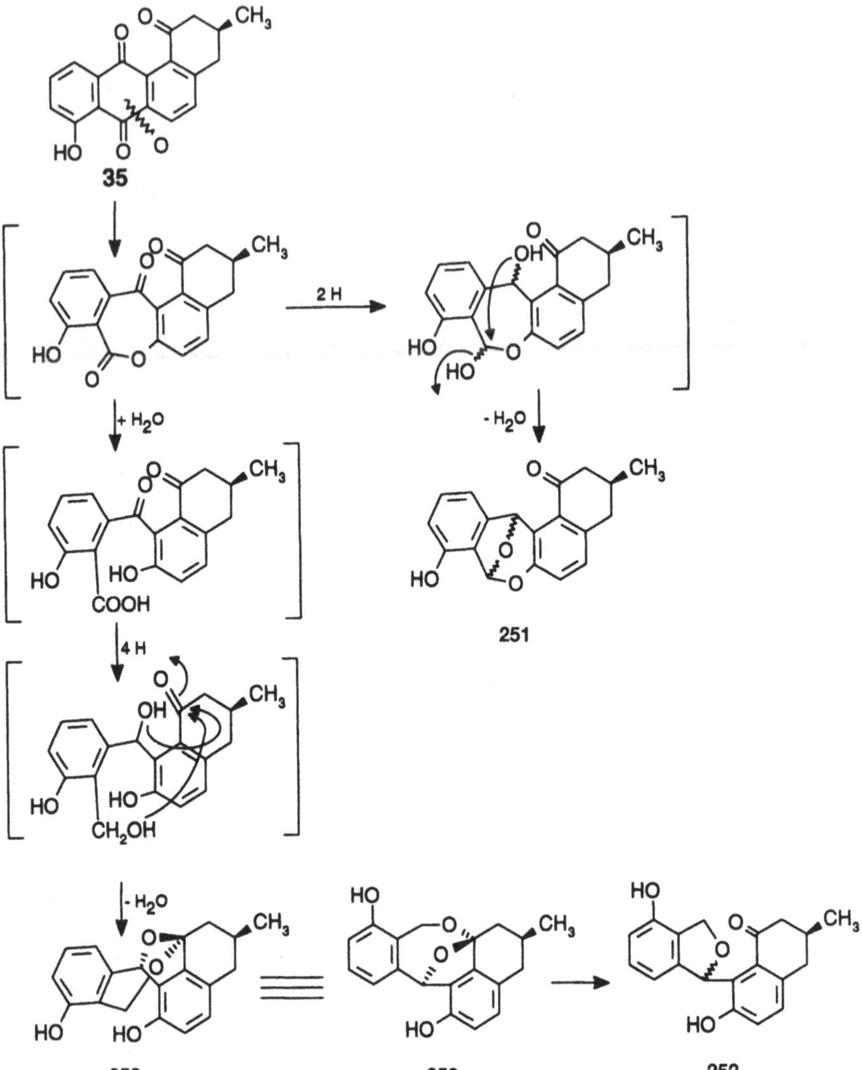

Scheme 67. Rationale for the biosynthetic formation of emycins D (**251**), E (**252**) and E (**253**). The key biosynthetic step is an oxidative cleavage of the 6a,7-bond of an angucyclinone, presumably ochromycinone (**35**)

ravidomycin (**293**), the chrysomycins A (**294**), and B (**295**), and thus BE-12406 A (**296**) and B (**297**, see Table 15), are all thought [226, 229, 230] to derive from a decaketide folded in the typical angucyclinone manner.

Similarly as shown for the kinamycins and jadomycins, an oxidative C-C bond cleavage initiates excision of a single carbon, and further rearrangement leads to the typical lactone structures of the gilvocarcin class. Although an angucyclinone intermediate has never been discussed for the biosynthesis of any of the gilvocarcins, and was – of course – never found, the proven biosyn-

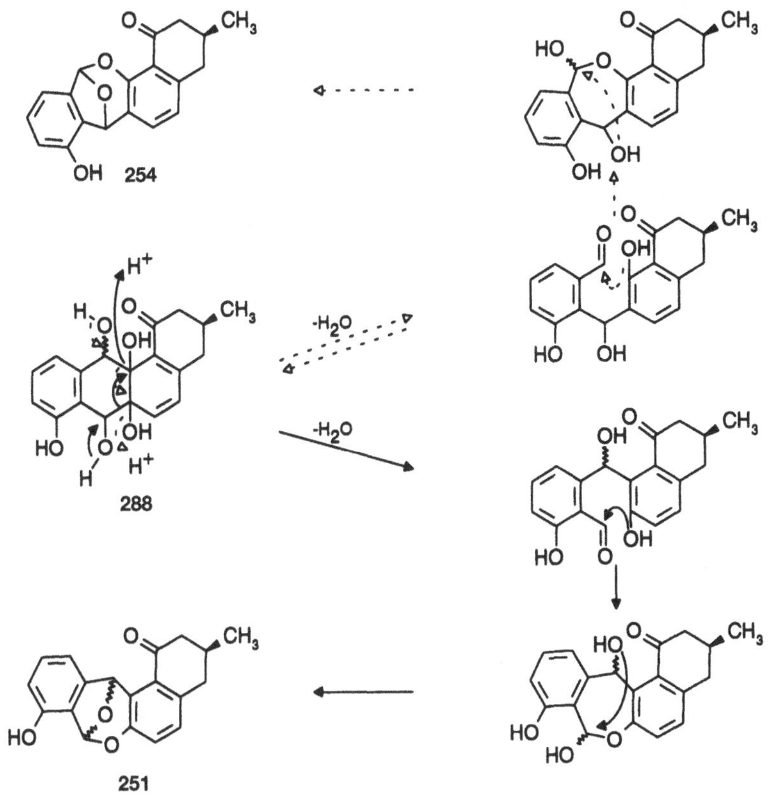

Scheme 68. Alternative rationale for the formation of emycins D (**251**) and G (**254**): 1,3-diol cleavage of emycin H analog **288**

thetic relationship of the kinamycins and jadomycins to the angucycline group antibiotics makes a biosynthetic affiliation of the gilvocarcin group very plausible (Scheme 69). The biosynthetic starter unit found in the gilvocarcin group varies between acetate and propionate.

Despite their interesting effects on stuctural diversity, the oxygenases involved in the biosyntheses of polyketides are not well studied. One exception is tetracenomycin A_2 oxygenase, which is coded by the gene *tcmG* and converts tetracenomycin A_2 into tetracenomycin C [242–244]. The tetracenomycin A_2 oxygenase, most likely a dioxygenase, is a monomeric protein in solution and contains one mole of noncovalently bound FAD (flavin adenine dinucleotide) [242]. Although in one case (mentioned above) an effect of a cytochrome P_{450} inhibitor could be observed [159], the majority of these interesting oxygenases seem not to be cytochrome P_{450}-dependent. For instance, we have never successfully changed the metabolite spectrum of any *Streptomyces* sp. using various P-450 inhibitors, as opposed to several successful examples with fungi. Also, the cytochrome P-450 oxygenase involved in the biosynthesis of PD 116198 (**218**) seems to only be responsible for the introduction of a single oxygen, namely, the one at C-2 [159].

Table 15. The gilvocarcin family of antibiotics

	R^1	R^2	R^3	R^4
289:	α-L-Fuc	CH_3	CH_3	CH_3
290:	α-L-Fuc	CH_2CH_3	CH_3	CH_3
291:	α-L-Fuc	$CH=CH_2$	CH_3	CH_3
292:	H	$CH=CH_2$	CH_3	CH_3
293:	α-D-AmSug	$CH=CH_2$	CH_3	CH_3
294:	β-L-MeSug	$CH=CH_2$	CH_3	CH_3
295:	β-L-MeSug	CH_3	CH_3	CH_3
296:	H	CH_3	CH_3	α-L-Rha
297:	H	CH_3	H	α-L-Rha

The observations that angucycline group antibiotic biosyntheses were stimulated by unspecific genetic modification procedures, as found for the antibiotic 34–1 (**207**) [146], product Tü 96 OR (**273**) of the lysolipin producer mutant [188], and for landomycin E (presumably **284**) [224], are unexpected and intriguing. They seem to indicate that the biosynthetic genes responsible for the formation of the tetracyclic ring system of the angucyclinones (and for further steps) belong to an evolutionarily old, sometimes silent, biosynthetic pathway, whose meaning for the producing organisms remains unclear. As a less romantic alternative, an involvement of angucycline biosynthetic pathways in the formation of spore pigments (e.g., for streptomycetes) may be assumed. Finally, as mentioned above in context with Tü 96 OR (**273**), which may have arisen from a mutation effect on the CLF, the folding leading to the angucycline frame may widely occur spontaneously.

As shown for other classes of polyketides, the genes involved in the biosynthesis of angucyclines may be very useful in the modern approach to creating hydrid natural products or "unnatural" natural products, or even in approaches aiming at combinatorial biosynthesis [190, 191, 194, 202b, 245–250]. For the creation of novel genetically engineered tetracenomycins, genes of the urdamycin producer *Streptomyces fradiae* Tü 2717, in particular those responsible for the institution of post-polyketide modifying enzymes, have already proven to be useful [217].

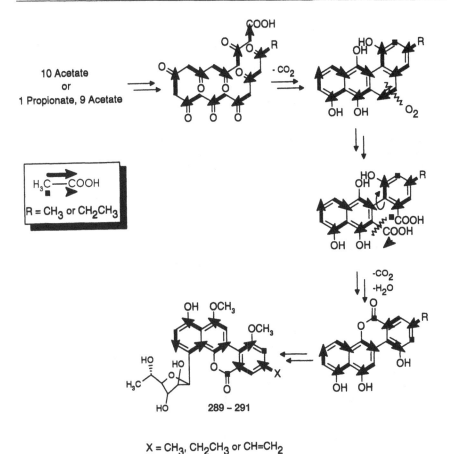

Scheme 69. Biosynthetic formation of the gilvocarcins via an angucyclinone. As in Scheme 66, the key biosynthetic step is an oxidative 5,6-bond cleavage

11
References

1. Begley S (1994) Newsweek March 28:39
2. Drautz H, Zähner H, Rohr J, Zeeck A (1986) J Antibiot 39:1657
3. Rohr J, Thiericke R (1992) Nat Prod Rep 9:103
4. Thomson RH (1987) Naturally Occurring Quinones III. Recent Advances. Chapman and Hall, London
5. Oka M, Konishi M, Oki T, Ohashi M (1990) Tetrahedron Lett 31:7473
6. Oka M, Kamel H, Hamagishi Y, Tomita K, Miyaki T, Konishi M, Oki T (1990) J Antibiot 43:967
7. Kunstmann MP, Mitscher LA (1966) J Org Chem 31:2920
8. Dann M, Lefemine DV, Barbatschi F, Shu P, Kunstmann MP, Mitscher LA, Bohonos N (1965) Antimicrob Agents Chemother 1965:832
9. Ayukawa S, Takeuchi T, Sezaki M, Hara T, Umezawa H, Nagatsu T (1968) J Antibiot 21:350
10. Sato Y, Gould SJ (1986) J Am Chem Soc 108:4625
11. Rohr JA, Zeeck, Floss HG (1988) J Antibiot 41:126

12. Rohr J, Zeeck A (1987) J Antibiot 40:459
13. Ohta K, Mizuta E, Okazaki H, Kishi T (1984) Chem Pharm Bull 32:4350
14. Kawashima A, Yoshimura Y, Goto J, Nakaike S, Mizutani T, Handa K, Omura S (1988) J Antibiot 41:1913
15. Sawada Y, Numata KI, Murakami T, Tanimichi H, Yamamoto S, Oki T (1990) J Antibiot 43:715
16. Kakushima M, Nishio M, Numata KI, Konishi M, Oki T (1990) J Antibiot 43:1028
17. Tsunakawa M, Nishio M, Ohkuma H, Tsuno T, Konishi M, Naito T, Oki T, Kawaguchi H (1989) J Org Chem 54:2532
18. Brown PM, Thomson RH (1976) J Chem Soc, Perkin Trans 1 1976:997
19. Jacobsen N, Torsell K (1973) Acta Chem Scand 27:3211
20. Sutherland JK, Towers P (1981) J Chem Soc, Chem Commun 1981:740
21. Krohn K (1990) Tetrahedron 46:291
22. Krohn K, Müller U, Priyono W, Sarstedt B, Stoffregen A (1984) Liebigs Ann Chem 1984:306
23. Krohn K, Dröge W, Hintze F (1995) Annal Quim 91:388
24. Arsenault G, Broadbent AD, Hutten-Czapski P (1983) J Chem Soc, Chem Commun 1983:437
25. Ashcroft AE, Davies DT, Sutherland JK (1984) Tetrahedron 40:4579
26. Ahmed Z, Cava MP (1981) Tetrahedron Lett 22:5239
27. Cava MP, Ahmed Z, Benfaremo N, Murphy Jr. RA, O'Malley GJ (1984) Tetrahedron 40:4767
28. Krohn K, Baltus W (1986) Synthesis 1986:942
29. Krohn K, Priyono W (1986) Angew Chem 98:338
30. Krohn K (1979) J Chem Res (S) 1979:318
31. Kraus GA, Wu Y (1991) Tetrahedron Lett 32:3803
32. Kraus GA, Wu Y (1995) Annal Quim 91:393
33. Fumagalli SE, Eugster CH (1971) Helv Chim Acta 54:959
34. Marschalk C, Koenig F, Ouroussoff N (1936) Bull Soc Chim F 3:1545
35. Krohn K, Sarsted B (1983) Angew Chem 95:897
36. Krohn K, Baltus W (1982) Liebigs Ann Chem 1982:1579
37. Krohn K, Ballwanz F, Baltus W (1993) Liebigs Ann Chem 1993:911
38. Lakshmikantham MV, Ravichandran K, Gosciniak DJ, Cava MP (1985) Tetrahedron Lett 26:4703
39. Krohn K, Khanbabaee K, Flörke U, Jones PG, Chrapkowski A (1994) Liebigs Ann Chem 1994:471
40. Gerken M, Blank S, Kolar C, Hermentin P (1989) J Carbohydr Chem 8:247
41. Yates P, Mackay AC, Garneau FX (1968) Tetrahedron Lett 1968:5389
42. Garcia-Garibay MA, Gamarnik A, Pang L, Jenks WS (1994) J Am Chem Soc 116:12095
43. Bowie JH, Johnson AW (1967) Tetrahedron Lett 1967:1449
44. Maehr H, Liu C-M, Liu M, Perrotta A, Smallheer JM, Williams T, Blount JF (1982) J Antibiot 35:1627
45. Katsuura K, Snieckus V (1985) Tetrahedron Lett 26:9
46. Katsuura K, Snieckus V (1987) Can J Chem 65:124
47. Valderrama JA, Araya-Maturana R, González MF, Tapia R, Fariña F, Paredes MC (1991) J Chem Soc, Perkin Trans 1 1991:555
48. Uemura M, Take K, Hayashi Y (1983) J Chem Soc, Chem Commun 1983:858
49. Uemura M, Take K, Isobe K, Minami T, Hayashi Y (1985) Tetrahedron 41:5771
50. Carothers WH, Coffman DD (1932) J Am Chem Soc 54:4071
51. Tomaszewski JE, Manning WB, Muschik GM (1977) Tetrahedron Lett 1977:971
52. Manning WB (1981) Tetrahedron Lett 22:1571
53. Guingant A, Barreto MM (1987) Tetrahedron Lett 28:3107
54. Gould SJ, Cheng X-C, Melville C (1994) J Am Chem Soc 116:1800
55. Krohn K (1986) Angew Chem 98:788
56. Stowell JC (1979) Carbanions in Organic Synthesis. Wiley Interscience, New York, p 37
57. Valderrama JA, Pessoa-Mahana CD, Tapia R (1994) J Chem Soc, Perkin Trans 1 1994:3521
58. Matsumoto T, Sohma T, Yamaguchi H, Kurata S, Suzuki K (1995) Synlett 1995:263
59. Matsumoto T, Sohma T, Yamaguchi H, Kurata s, Suzuki K (1995) Tetrahedron 51:7347

60. Tochtermann W, Malchow A, Timm H (1978) Chem Ber 111:1233
61. Bridson JN, Bennett SM, Butler G (1980) J Chem Soc, Chem Commun 1980:413
62. Larsen DS, O'Shea MD (1993) Tetrahedron Lett 34:1373
63. Larsen DS, O'Shea MD (1993) Tetrahedron Lett 34:3769
64. Larsen DS, OÇShea MD (1995) J Chem Soc, Chem Commun 1995:1019
65. Sezaki M, Kondo S, Maeda K, Umezawa H (1970) Tetrahedron 26:5171
66. Larsen DS, OÇShea MD, Brooker S (1996) J Chem Soc, Chem Commun 1996:203
67. Kelly TR, Whiting A, Chandrakumar NS (1986) J Am Chem Soc 108:3510
68. Krohn K, Khanbabaee K (1994) Angew Chem 106:100
69. Fleming I, Henning R, Plaut H (1984) J Chem Soc, Chem Commun 1984:29
70. Still WC (1976) J Org Chem 41:3063
71. Hudrlik PF, Hudrlik AM, Nagendrappa G, Yimenu T, Zellers ET, Chin E (1980) J Am Chem Soc 102:6896
72. Chan TH, Nwe KT (1992) J Org Chem 57:6107
73. Krohn K, Khanbabaee K (1994) Liebigs Ann Chem 1994:1109
74. Krohn K, Khanbabaee K, Micheel J (1995) Liebigs Ann Chem 1995:1529
75. Micheel J (1995) Diploma thesis. Universität Paderborn
76. Matsuo G, Miki Y, Nakata M, Matsumura S, Toshima K (1996) J Chem Soc, Chem Commun 1996:225
77. Gilpin ML, Balchin J, Box SJ, Tyler JW (1989) J Antibiot 42:627
78. Krohn K (1980) Tetrahedron Lett 21:3557
79. Laatsch H (1985) Liebigs Ann Chem 1985:251
80. Boeckman Jr. RK, Delton MH, Dolak TM, Watanabe T, Glick MD (1979) J Org Chem 44:4396
81. Rozeboom MD, Tegmo-Larsson I-M, Houk KN (1981) J Org Chem 46:2338
82. Krohn K, Khanbabaee K, Jones PG (1995) Liebigs Ann Chem 1995:1981
83. Kim K, Reibenspies J, Sulikowski G (1992) J Org Chem 57:5557
84. Kim K, Guo Y, Sulikowski, G. A. (1995) J Org Chem 60:6866
85. Boyd VA, Reibenspies J, Sulikowski GA (1995) Tetrahedron Lett 36:4001
86. Krohn K, Böker N, Freund C (1996) J Org Chem 1996 (in preparation)
87. Boyd VA, Sulikowski GA (1995) J Am Chem Soc 117:8472
88. Billen G, Scholl KU, Stroech KD, Steglich W (1988) In: HEJ Atta-ur-Rahman (eds) Natural Products Chemistry. Springer, Heidelberg, p 305
89. Kim K, Sulikowski, G. A. (1995) Angew Chem 107:2587
90. Sasaki T, Yoshida J, Itoh M, Gomi S, Shomura T, Sezaki M (1988) J Antibiot 41:835
91. Sasaki T, Gomi S, Sezaki M, Takeuchi Y, Kodoma Y, Kawamura K (1988) J Antibiot 41:843
92. Kraus G, Zhao G (1995) Synlett 1995:541
93. Kraus G, Zhao G (1996) J Org Chem 61:2770
94. Nicolas TE, Franck RW (1995) J Org Chem 60:6904
95. Rohr J (1992) J Org Chem 57:5217
96. Gould SJ, Halley KA (1991) J Am Chem Soc 113:5092
97. Yamaguchi M, Okuma T, Horiguchi A, Ikeura C, Minami T (1992) J Org Chem 57:1647
98. Yamaguchi M, Hasebe K, Higashi H, Uchida M, Irie A, Minami T (1990) J Org Chem 55:1611
99. Harris TM, Harris CM, Hindley KB (1974) Fortschr Chem Org Naturstoff 31:217
100. Krohn K, Roemer E, Top M, Wagner C (1993) Angew Chem 105:1220
101. Krohn K, Schäfer G (1996) Liebigs Ann Chem 1996:265
102. Krohn K, Roemer E, Top M (1996) Liebigs Ann Chem 1996:271
103. Grunwell JR, Heinzman SW (1980) Tetrahedron Lett 21:4305
104. Wurm G, Geres U (1984) Arch Pharm (Weinheim) 317:606
105. Wurm G, Gurka H-J (1986) Arch Pharm (Weinheim) 319:190
106. Stille JK, Groh BL (1987) J Am Chem soc 109:813
107. Stille JK (1986) Angew Chem, Int Ed Engl 25:508
108. Tamayo N, Echavarren AM, Paredes MC, Fariña F, Noheda P (1990) Tetrahedron Lett 31:5189

109. Mata EG, Mascaretti OA (1988) Tetrahedron Lett 29:6893
110. Krohn K (1989) Building blocks for the total synthesis of anthracyclinones. In: W Herz; H Grisebach; GW Kirby; Ch Tamm (eds) Prog Chem Org Nat Prod. Springer, Wien, New York, p 37
111. Chuang C-P, Wang S-F (1994) Tetrahedron Lett 35:4365
112. Gordon DM, Danishefsky SJ, Schulte GK (1992) J Org Chem 57:7052
113. Dötz KH, Popall M, Müller G (1985) J Organomet Chem 291:C1
114. Wulff WD, Xu Y-C (1988) J Am Chem Soc 110:2312
115. Oki T, Konishi M, Tomatsu K, Tomita K, Saitoh KI, Tsunakawa M, Nishio M, Miyaki T, Kawaguchi H (1988) J Antibiot 41:1701
116. Tomita K, Nishio M, Saitoh K, Yamamoto H, Hoshino Y, Ohkuma H, Konishi M, Miyaki T, Oki T (1990) J Antibiot 43:755
117. Oki T, Kakushima M, Nishio M, Kamel H, Hirano M, Swada Y, Konishi M (1990) J Antibiot 43:1230
118. Oki T, Tenmyo O, Hirano M, Tomatsu K, Kamai H (1990) J Antibiot 43:763
119. Kakushima M, Masuyoshi S, Hirano M, Shinoda M, Ohta A, Kamei H, Oki T (1991) Antimicrob Agents Chemother 35:2185
120. Oki T, Kakushima M, Hirano M, Takahashi A, Ohta A, Masuyoshi S, Hatori M, Kamei H (1992) J Antibiot 45:1512
121. Oki T (1991) Recent progress in antifungal chemotherapy. In: H Yamaguchi, GS Kobayashi, H Takahashi (eds) Marcel Dekker; New York, p 381
122. Gerber NN, Lechevalier MP (1984) Can J Chem 62:2818
123. Rickards RW (1989) J Antibiot 42:336
124. Kelly TR, Li Q, Bhushan V (1990) Tetrahedron Lett 31:161
125. Kelly TR, Xu W, Ma Z, Li Q, Bushan V (1993) J Am Chem Soc 115:5843
126. Hauser FM, Caringal Y (1990) J Org Chem 55:555
127. Yamaguchi M, Okuma T, Ikeura C, Minami T (1992) J Chem Soc, Chem Commun 1992:434
128. Daves GD, Jr (1990) Acc Chem Res 23:201
129. Jamarillo C, Knapp S (1993) Synthesis 1993:
130. Suzuki K, Matsumoto T (1993) Total synthesis of Aryl C-glycoside antibiotics. In: G Lukacs (ed) Recent progress in the chemical synthesis of antibiotics and related microbial products. Springer, Berlin, Heidelberg, New York, p 353
131. Kwok D-I, Outten RO, Huhn R, Daves GD, Jr. (1988) J Org Chem 53:5359
132. Outten RA, Daves GD (1989) J Org Chem 54:29
133. Matsumoto T, Hosoya T, Suzuki K (1991) Synlett 1991:709
134. Matsumoto T, Hosoya T, Suzuki K (1992) J Am Chem Soc 114:3568
135. Tius MC, Gu X, Gomez-Galeno G (1990) J Am Chem Soc 112:8188
136. Tius MA, Gomez-Galeno J, Gu X-q, Zaidi JH (1991) J Am Chem Soc 113:5775
137. Boyd VA, Drake BE, Sulikowski GA (1993) J Org Chem 58:3191
138. Arhart RJ, Martin JC (1972) J Am Chem Soc 94:5003
139. Andrews FL, Larsen DS (1994) Tetrahedron Lett 35:8693
140. Brinkman LC, Ley FR, Seaton PJ (1993) J Nat Prod 56:374
141. Okabe T, Ogino H, Suzuki H, Suda AOH (06.11.92) Jpn Kokai Tokkyo Koho 04,316,492 [92,316,492]; CA 118:167614v
142. Hayakawa Y, Ha S-C, Kim YJ, Furihata K, Seto H (1991) J Antibiot 44:1179
143. Seto H, Hayakawa Y, Shimazu A (23.10.1992) Jpn. Kokai Tokkyo Koho 04,300,849 [92,300, 849]; CA 118:146245k
144. Fujioka K, Furihata K, Shimazu A, Hayakawa Y, Seto H (1991) J Antibiot 44:1025
145. Seto H, Hayakawa Y, Shimazu A, Furuhata K, Fujioka K (02.06.1992) JP 04,159,291 [92, 159,291] CA 117: 232216n
146. Lazhko EI, Novozhenov MY, Malanicheva IA (1992) Bioorg Khim 18:1519
147. Weber S, Zolke C, Rohr J, Beale JM (1994) J Org Chem 59:4211
148. Henkel T, Rohr J, Beale JM, Schwenen L (1990) J Antibiot 43:492
149. Maul C, Zerlin M, Wohlert S-E, Rohr J (1996) unpublished results
150. Kanamura T, Nozaki Y, Muroi M (29.11.1990) JP 02,289,532 [90, 289, 532] CA 115:47759n

151. Igarashi M, Sasao C, Yoshida A, Naganawa H, Hamada M, Takeuchi T (1995) J Antibiot 48:335
152. Senda S, Seto H (27.12.1989) Eur Pat Appl EP 435,221 CA 115:112826y
153. Ritzau M (1992) Dissertation, Universität Göttingen
154. Tanaka Y, Sugoh M, Yoshida H, Arai N, Shiomi K, Matsumoto A, Takahashi Y, Omura S (1995) J Antibiot 48:1525
155. Gerlitz M, Udvarnoki G, Rohr J (1995) GIT Fachz Lab 39:888
156. Gerlitz M, Udvarnoki G, Rohr J (1995) Angew Chem Int Ed Engl 34:1617
157. Gerlitz M (1995) Dissertation, Universität Göttingen, Cuvillier, Göttingen
158. (a) Phife DW, Patton RW, Berrie RL, Yarborough R, Puar MS, Patel M, Bishop WR, Coval SJ (1995) Tetrahedron Lett 36:6995 and (b) Wege D (1996) Aust J Chem 49:669
159. Gould SJ, Cheng XC (1994) J Org Chem 59:400
160. Miyata S, Ohhata N, H, Murai, Masui Y, Ezaki M, Takase S, Nishikawa M, Kiyoto S, Okuhara M, Kohsaka M (1992) J Antibiot 45:1029
161. Miyata S, Hashimoto M, Fujie K, Shouho M, Sogabe K, Kiyoto S, Okuhara M, Kohsaka M (1992) J Antibiot 45:1041
162. Oohata N, Motoaki N, Kiyoto S, Takase S, Hemmi K, Murai H, Okuhara M (30.6.1989). Eur. Pat Appl EP 405,421 CA 115:157125k
163. Etoh H, Iguchi M, Nagasawa T, Tani Y, Yamada H, Fukami H (1987) Agric Biol Chem 51:1819
164. Rohr J, Schönewolf M, Udvarnoki G, Eckardt K, Schumann G, Wagner C, Beale JM, Sorey SD (1993) J Org Chem 58:2547
165. Bindseil KU, Hug P, Peter HH, Petersen F, Roggo BE (1995) J Antibiot 48:457
166. Bethe B (1994), Dissertation, Universität Göttingen, Cuviller, Göttingen
167. Chu M, Yarborough R, Schwartz J, Patel MG, Horan AC, Gullo VP, Das PR, Puar MS (1993) J Antibiot 46:861
168. Okabe T, Funaishi K, Hagiwara M, Kawamura K, Suda H, Sato F, Okanishi M (03.12.1990) JP 02,291,287 [90, 291, 287] CA 115: 69926s
169. Imamura N, Kakinuma K, Ikekawa N, Tanaka H, Omura S (1981) J Antibiot 34:1517
170. Kawamura N, Sawa R, Takahashi Y, Sawa T, Kinoshita N, Naganawa H, Hamada M, Takeuchi T (1995) J Antibiot 48:1521
171. Sawa R, Matsuda N, Uchida T, Ikeda T, Sawa T, Naganawa H, Hamada M, Takeuchi T (1991) J Antibiot 44:396
172. Ojiri K, Suda H, Okura A, Kawamura K, Okanishi M (08.02.91) JP 0330,687 [91, 30, 687] CA 115:157135p
173. Kawashima A, Maejima A, Miyoshi T, Tamai M, Hanada K, Omura S (22.01.1990) JP 03,220,196 [91, 220, 196] CA 116:104492s
174. Müller H, Fugmann B, Weber K, Schmidt D (1991) Dechema/GDCh-. Symposium: „Neue niedermolekulare Naturstoffe und Perspektiven für ihre Anwendung", Irsee/Bayern, 13.2.–15.2.1991
175. Kawashima A, Hamaguchi T, Amaka T, Hanada K (1992) JP 04,178,379, CA 118:21058v
176. (1991) Umezawa H, Takeuchi T, Sawa T, Hamada M, Naganawa H, Uchida T, Imoto M (1991) EP 191-399-B
177. Mithani S, Weeratunga G, Taylor NJ, Dmitrienko GI (1994) J Am Chem Soc 116:2209
178. Gould SJ, Tamayo N, Melville CR, Cone MC (1994) J Am Chem Soc 116:2207
179. Seaton PJ, Gould SJ (1987) J Am Chem Soc 109:5282
180. Smitka TA, Bonjouklian R, Perun J,TJ, Hunt AH, Foster RS, Mynderse JS, Yao RC (1992) J Antibiot 45:581
181. Lin H-C, Chang S-C, Wang N-L, Chang l-R (1994) J Antibiot 47:675
182. Young J-J, Ho S-N, Ju W-M, Chang l-R (1994) J Antibiot 47:681
183. (a) Cone MC, Melville CR, Gore MP, Gould SJ (1993) J Org Chem 58:1058 and (b) Gould SJ, Melville CR (1995) Bioorganic Med Chem Lett 6:51
184. Aoyama T, Zhao W, Kojima F, Muraoka, Y, Naganawa H, Takeuchi T, Aoyagi, T (1993) J Antibiot 46:1471
185. Seto H, Hayakawa Y, Shimazu A (1993) JP 05,85,998 [93 85, 998], CA 119:70550d

186. Shin-ya K, Furihata K, Teshima Y (1992) Tetrahedron Lett 33:7025
187. Volkmann C, Rössner E, Metzler M, Zähner H, Zeeck A (1995) Liebigs Ann Chem 1995:1169
188. Weissinger M, Bockholt H, Rohr J (1996) unpublished
189. Bockholt H, Udvarnoki G, Rohr J, Mocek U, Beale JM, Floss HG (1994) J Org Chem 59:2064
190. McDaniel R, Ebert-Khosla S, Hopwood DA, Khosla C (1993) Science 262:1546
191. McDaniel R, Ebert-Khosla S, Hopwood DA, Khosla C (1995) Nature 375:549
192. McDaniel R, Ebert-Khosla S, Hopwood DA, Khosla C (1993) J Am Chem Soc 115:11671
193. Shen B, Summers RG, Wendt-Pienkowski E, Hutchinson CR (1995) J Am Chem Soc 117:6811
194. Rohr J (1995) Angew Chem Int Ed Engl 34:881
195. Ayer SW, McInnes AG, Thibault P, Walter JA, Doull JL, Parnell T, Vining LC (1991) Tetrahedron Lett 32:6301
196. Doull JL, Ayer SW, Singh AK, Thibault P (1993) J Antibiot 46:869
197. Han L, Yang K, Ramalingam E, Mosher RH, Vining LC (1994) Microbiology 140:3379
198. Seaton PJ, Gould SJ (1988) J Am Chem Soc 110:5912
199. Cone MC, Hassan AM, Gore MP, Gould SJ, Borders DB, Alluri MR (1994) J Org Chem 59:1923
200. Yang K, Han L, Ayer SW, Vining LC (1996) Microbiology 142:123
201. O'Hagan D (1991) The polyketide metabolites. Ellis Horwood, New York
202. (a) O'Hagan D (1995) Nat Prod Rep 12:1 and (b) Hutchinson CR, Fujii I (1995) Annu Rev Microbiol 49:201
203. Tsuda M, Sato H, Tanaka Y, Yazawa K, Mikami Y, Sasaki T, Kobayashi J (1996) J Chem Soc Perkin Trans 1:773
204. Udvarnoki G, Henkel T, Machinek R, Rohr J (1992) J Org Chem 57:1274
205. Gould SJ, Cheng X-C, Halley KR (1992) J Am Chem Soc 114:10066
206. Gould SJ, Cheng X-C (1993) Tetrahedron 49:11135
207. Weibbach U (1996) Diplomarbeit, Universität Göttingen
208. Bechthold A (1996) private communication
209. Donadio S, Staver MJ, McAlpine JB, Swanson SJ, Katz L (1991) Science 252:675
210. Cortes J, Haydock SF, Roberts GA, Bevitt DB, Leadlay PF (1990) Nature 348:176
211. Leadlay PF, Staunton J, Aparicio JF, Bevitt DJ, Caffrey P, Cortes J, Marsden A, Roberts GA (1993) Biochem Soc Trans 21:218
212. Anderson JA, Lin B-K, Williams H-J, Scott AI (1988) J Am Chem Soc 110:1623
213. Erickson WR, Gould SJ (1985) J Am Chem Soc 107:5831
214. Gould SJ, Cheng X-C (1993) Tetrahedron 49:11135
215. Decker H, Haag S (1995) J Bacteriol 177:6126
216. Decker H, Rohr J, Motamedi H, Zähner H, Hutchinson CR (1995) Gene 166:121
217. Decker H, Haag S, Udvarnoki G, Rohr J (1995) Angew Chem Int Ed Engl 34:1107
218. Decker H, Gaisser S, Pelzer S, Schneider P, Westrich L, Wohlleben W, Bechthold A (1996) FEMS Lett, in press
219. Beninga C, Wohlert S-E, Rohr J (1996) unpublished results
220. Beninga C (1994) Diplomarbeit, Universität Göttingen
221. Wohlert S-E (1994) Diplomarbeit, Universität Göttingen
222. Polishchuk L, Stefanishin EE, Dekhtyarenko TD, Sten'ko AS, Zaverukha VB, Matselyukh BP (1987) Mikrobiol Z (Ukr) 49:24
223. Zaverukha VB, Polevoda BV, Polishchuk LV, Matselyukh BP (1992) Mikrobiol Z (Ukr) 54:30
224. Matselyukh B, Polishchuk L, Weibbach U, Wohlert S-E, Rohr J (1996) J Antibiot, in preparation
225. Cone MC, Melville CR, Carney JR, Gore MP, Gould SJ (1995) Tetrahedron 51:3095
226. Carter GT, Fantini AA, James JC, Borders DB, White RJ (1984) Tetrahedron Lett 25:255
227. Elespuru RK, Gonda SK (1984) Science 223:69
228. Morimoto M, Okubo S, Tomita F, Marumo H (1981) J Antibiot 34:701

229. Takahashi K, Tomita F (1983) J Antibiot 36:1531
230. Carter GT, Fantini AA, James JC, Borders DB, White RJ (1985) J Antibiot 38:242
231. Misra R, Tritch HR, Pandey RC (1985) J Antibiot 38:1280
232. Keyes RF, Kingston DGI (1989) J Org Chem 54:6127
233. Eguchi T, Li H-Y, Kazami J-I, Kakinuma K, Otake N (1990) J Antibiot 43:1077
234. Balitz DM, O'Herron FA, Bush J, Vyas DM, Nettleton DE, Grulich RE, Bradner WT, Doyle TW, Arnold E, Clardy J (1981) J Antibiot 34:1544
235. Hatano K, Higashide E, Shibata M, Kameda Y, Horii S (19801) Agric Biol Chem 44:1157
236. McGee LR, Misra R (1990) J Am Chem Soc 112:2386
237. Nakajima S, Kojiri K, Suda H, Okanishi M (1991) J Antibiot 44:1061
238. Kojiri K, Arakawa H, Satoh F, Kawamura K, Okura A, Suda H, Okanishi M (1991) J Antibiot 44:1054
239. Kikuchi O, Eguchi T, Kakinuma K, Koezuka Y, Shindo K, Otake N (1993) J Antibiot 46:985
240. Hosoya T, Takashiro E, Matsumoto T, Suzuki K (1994) J Am Chem Soc 116:1004
241. Farr RN, Kwok DI, Daves J,GD (1992) J Org Chem 57:2093
242. Shen B, Hutchinson CR (1994) J Bacteriol 269:30726
243. Decker H, Motamedi H, Hutchinson CR (1993) J Bacteriol 175:3876
244. Udvarnoki G, Wagner C, Machinek R, Rohr J (1995) Angew Chem Int Ed Engl 34:565
245. Cane DE (1994) Science 263:338
246. Piepersberg W (1994) Critical Rev Biotechnol 14:251
247. Tsoi CJ, Khosla C (1995) Chemistry & Biology 2:355
248. Katz L, Donadio S (1993) Annu Rev Microbiol 47:875
249. Fu H, Hopwood DA, Khosla C (1994) Chemistry & Biology 1:205
250. Rouhi M (1995) C&EN 1995:9

Printing: Saladruck, Berlin
Binding: H. Stürtz AG, Würzburg